T0136818

A Scientific Peak

How Boulder Became a World Center for
Space and Atmospheric Science

Joseph P. Bassi

American Meteorological Society

Cover photo by Charles Pfeil

Published by the American Meteorological Society
45 Beacon Street, Boston, Massachusetts 02108

The mission of the American Meteorological Society is to advance the atmospheric and related sciences, technologies, applications, and services for the benefit of society. Founded in 1919, the AMS has a membership of more than 13,000 and represents the premier scientific and professional society serving the atmospheric and related sciences. Additional information regarding society activities and membership can be found at www.ametsoc.org.

Library of Congress Cataloging-in-Publication Data

Bassi, Joseph P., 1952–
 A scientific peak : how Boulder became a world center for space and atmospheric science / Joseph P. Bassi.
     pages cm
 Includes bibliographical references and index.
  ISBN 978-1-935704-85-0 (pbk.)
  1. Science—Colorado—Boulder. 2. Technology—Colorado—Boulder. 3. Space sciences—Colorado—Boulder. 4. Atmospheric physics—Research—Colorado—Boulder. 5. Boulder (Colo.) I. Title.
  Q127.U6B277 2015
  507.2'078863—dc23
                    2015029849

SUSTAINABLE
FORESTRY
INITIATIVE

To Marianne, my dear wife

Contents

List of Abbreviations and Acronyms   vii
List of Archives Consulted   xi
Foreword   xiii
Acknowledgments   xvii

1   Introduction   1
2   Sun–Earth Science Arrives in Colorado   17
3   From Leadville to Boulder   35
4   A Scientific Peak Begins to Develop in Boulder   59
5   "Nothing but a Fundraiser"   77
6   Global Science in One Place   105
7   An Atmosphere of Change   125
8   NCAR and Boulder's Entry into the "Environmental Era"   147
9   Conclusion   169

Images   179
Notes   195
Index   237

List of Abbreviations and Acronyms

AAF: Army Air Forces
ACWC: Advisory Committee on Weather Control
AGU: American Geophysical Union
AHC: Alan H. Shapley Collection
AIP: American Institute of Physics
AUI: Associated Universities, Inc.
BDC: *Boulder Daily Camera*
CHP: Center for the History of Physics
CMC: Climax Molybdenum Company
COM: National Academy of Sciences Committee on Meteorology
CRPL: Central Radio Propagation Laboratory
CSAGI: Comité Spécial de l'Année Géophysique Internationale
CSO: Committee on Scientific Operations
CU: University of Colorado Boulder
CUA: University of Colorado Archives
DHM: Donald Howard Menzel
DOD: Department of Defense
DPL: Denver Public Library
DTM: Department of Terrestrial Magnetism
ESSA: Environmental Science Services Administration

HAO: High Altitude Observatory
HCO: Harvard College Observatory
HUA: Harvard University Archives
HUAC: House Un-American Activities Committee
HW: Harry Wexler Papers
IAU: International Astronomical Union
IBM: International Business Machines
IGY: International Geophysical Year
IGYWWA: International Geophysical Year World Warning Agency
IPY: International Polar Year
IRPL: Interservice Radio Propagation Laboratory
ISTR: Institute for Solar–Terrestrial Relations
IUGG: International Union of Geodesy and Geophysics
JILA: Joint Institute for Laboratory Astrophysics, University of Colorado
JSC: Johnson Space Center
LASP: Laboratory for Atmospheric and Space Physics, University of Colorado
LOC: Library of Congress
MIT: Massachusetts Institute of Technology
NARA: National Archives and Records Administration
NAS: National Academy of Sciences
NASA: National Aeronautics and Space Administration
NBS: National Bureau of Standards
NCAR: National Center for Atmospheric Research
NIAR: National Institute of Atmospheric Research
NIST: National Institute of Standards and Technology
NOAA: National Oceanic and Atmospheric Administration
NRC: National Research Council
NSB: National Science Board
NSF: National Science Foundation
OASDRE: Office of Assistant Secretary of Defense for Research and Engineering
OHP: Oral History Program
ONR: Office of Naval Research
PBC: Plan Boulder County, current acronym for former PLAN
PF: President's Office Files
PLAN: People's League for Action Now, Boulder
QN: Quigg Newton Papers
RCA: Radio Corporation of America

SAO: Smithsonian Astrophysical Observatory
SIO: Scripps Institution of Oceanography
SWI: Special World Interval
TWA: Trans World Airlines
UAL: Upper Air Laboratory
UCAR: University Committee on Atmospheric Research/University Corporation for Atmospheric Research
UCARA: University Corporation for Atmospheric Research Archives
UCLA: University of California, Los Angeles
URSI: Commission on the Ionosphere of the International Scientific Radio Union
USNC: United States National Committee
USWB: United States Weather Bureau
WDC: World Data Centers
WHOI: Woods Hole Oceanographic Institute
WOR: Walter Orr Roberts Collection
WWV: Call sign of National Institute of Standards and Technology radio station

List of Archives Consulted

American Institute of Physics/Center for the History of Physics, College Park, MD

Carnegie Library, Boulder, CO

Denver Public Library, Denver, CO

Harvard University Archives, Cambridge, MA

Library of Congress, Washington, D.C.

National Academy of Sciences, Washington, D.C.

National Archives and Records Administration, College Park, MD

Scripps Institution of Oceanography, La Jolla, CA

University Corporation for Atmospheric Research Archives/National Center for Atmospheric Research, Boulder, CO

University of Colorado Archives, Boulder, CO

Foreword

I remember the first time I laid eyes on Boulder. It was in the summer of 1985, and I had several interviews planned at what was then known as the Program for Regional Observing and Forecasting Services (PROFS). I hoped to land my first professional job there, anticipating completion of my master's degree in meteorology at Penn State the following year. Being a lifelong New Englander, I might have had some trepidation relocating 2000 miles to the west to start my career, and I do remember feeling nervous excitement as I traveled the interstate north from Denver that first time. But any misgivings I might have had vanished as I crested a small incline and drank in the sight of Boulder Valley, which at that moment was entirely contained within one of the most vivid rainbows I have ever seen, stretching from the Flatirons to the west to the plains in the east. I don't know if I believe in omens, but I certainly did that day. My job at PROFS started in April 1986, and for the next three and a half years I had the privilege of living, working, and playing in that enchanted place. It was the first of two jobs that would eventually lead me to my current position at the American Meteorological Society a dozen years hence.

PROFS, part of the National Oceanic and Atmospheric Administration (NOAA) and forerunner to today's Forecast Systems Laboratory, certainly wasn't the only possible landing spot for an aspiring atmospheric scientist at that time. Boulder had long been established as the preeminent center for atmospheric research, boasting a number of other prominent NOAA labs as well as the National Center for Atmospheric Research (NCAR). The University of Colorado at Boulder was also part of the picture, partnering with NOAA through the Cooperative Institute for Research in Environmental Sciences (CIRES). It is easy to see why my graduate school colleagues and I saw Boulder as a meteorological mecca—the apex of atmospheric sciences research. I was too busy starting my career there to wonder how that had all come to be. Why Colorado? Why Boulder? Why should I care? To be honest, I never gave it a second thought until I read this book, and I haven't stopped thinking about it since.

My first thought upon establishing myself in Boulder was how orderly it seemed to be. I was used to Boston, where the streets are tangled webs of old cow paths, one-way streets, and drivers who seem to constantly be on the edge of road rage. Then there's Boulder, with its sensible grid, streets that are clearly named, and dedicated bicycle lanes. Even the mountains add to the order, all to the west of town, making it instantly apparent even to the directionally challenged which way north, south, and east are aligned. Everything made sense, and it's clear that the city's planners had a logical rationale for Boulder's growth. But don't confuse the evolution of the town with the evolution of the town's intellectual resources. Joe Bassi quickly disavows us of any notion that Boulder was a planned "Science City." There was no manifest destiny at work here in that regard. To the contrary, one comes away from reading this book marveling at the staggering odds against Boulder's transformation from a "scientific Siberia" in the years leading up to World War II to a world center of knowledge production in less than three decades. There were any number of twists and turns along the way at which a single decision to direct funding or laboratories to another city would have scuttled everything. At these precarious moments, even though I knew the eventual outcome, I found myself actually *rooting* for Boulder and for the steady force behind its rise to scientific prominence, Walter O. Roberts.

Similarly, there is no way that the soul of a city can be planned. I found that there was, and still is, a Boulder state of mind. People are incredibly dedicated to their work, but somehow take their fun just as seriously. There is a palpable feeling that anything is possible and that everyone can and should make the world a better place. That, as much as the state-of-the art laboratories, attracts some of the best scientists to the foothills of the Rocky Mountains at some point in their careers. Indeed, most people who live and work in Boulder seem to be from somewhere else. Most folks who move to

Boulder never leave (and those few that do wish they had never left). But the intellectual and scientific capital generated there by the people employed in atmospheric science, geophysics, geology, and space sciences is exported to the rest of the world in a way that enriches scientific inquiry everywhere.

I love Boulder. It is a jewel of a city, replete not only with some of the best minds and scientific laboratories in the world, but also blessed with unparalleled natural beauty at its doorstep under skies that are somehow bluer than blue. Can't we just leave it at that and appreciate it for what it is? We could, but we would be selling ourselves short. While *A Scientific Peak* is not a cookbook for creating a glittering center of scientific inquiry and industry, and the recipe that made Boulder can never work again anywhere else in exactly the same way, history does inform the present and the future. Some of the ingredients of the metamorphosis we read about here are timeless and can be used to great advantage in the future. Among these are openness to the free exchange of scientific knowledge; persistence and determination in the face of inevitable obstacles; and most crucially, the ability to recognize and seize opportunities when they arise. A little luck never hurts either.

Kenneth F. Heideman
Director of Publications
American Meteorological Society

Acknowledgments

A project of this scope necessarily had many contributors. My dissertation advisor at the University of California, Santa Barbara, W. Patrick McCray, provided much useful guidance over the course of my doctoral studies, and I am most appreciative. Mike Osborne, also of UCSB, was a constant source of encouragement and ideas as I progressed on this work. The book itself began as a paper for Alice O'Connor's research seminar, and her enthusiasm for this project served over the years as an important component of its completion. To all, many thanks—especially for the perseverance to wade through the results of my two-fingered typing.

Diane Rabson, former archivist at the University Corporation for Atmospheric Research (UCAR), helped immeasurably as I researched events in Boulder and Walter Roberts's long career. Her knowledge of the UCAR collection, plus insights in how best to research my topic, saved me much wasted time and effort. Diane's welcoming attitude toward me (and all scholars) in no small way made research in Boulder a very positive experience for me as a maturing researcher. Kate Legg, now at UCAR Archives, offered me much assistance as well. Similarly, David Hays at the University of Colorado Boulder Archives offered many suggestions on how best to address both the Walter Orr Roberts papers and University of Colorado Boulder history. In addition, his "find" of the Senator "Big Ed" Jackson file at the CU Archives assisted this work greatly. Jane Odom and others at the NASA HQ History Office Archives also provided a most congenial place to work and made me feel almost part of the staff. I am most thankful to all of these great archivists who welcomed me so generously into their world, and to a myriad of other archivists that contributed to my work at the Carnegie Library in Boulder, the National Academy of Sciences, the American Institute of Physics, the Library of Congress, the National Archives and Records Administration, and the Scripps Institute.

An old saying in Washington, D.C., about human spaceflight is "No Bucks, no Buck Rodgers." Similarly, "no bucks, no PhD research," so I thank all who helped to fund my work. The National Science Foundation Integrative Graduate Education and Research Traineeship (IGERT) for Public Policy and Nuclear Threats (PPNT) provided the basic assistance over a four-year period needed for my studies. Administered through the Institute on Global Conflict and Cooperation (IGCC) at the University of California, San Diego, IGERT was essential to the completion of this book. IGCC was also an academic home for me, and I thank Professor Susan Shirk for setting up the IGERT PPNT program. The award of a Guggenheim Pre-Doctoral fellowship at the Smithsonian Institution National Air and Space Museum Department of Space History enabled me to live in Washington, D.C., for six months in support of this project. The American Institute of Physics also provided me a grant for travel to their excellent historical archive collection in College Park, Maryland.

Peter Westwick and Jim Fleming provided particular insight and encouragement as I tackled this complex project, and I am in their debt. The entire staff of the Department of Space History, National Air and Space Museum, provided me with a stimulating intellectual environment during my D.C. stay. Of course, my editor at AMS Books, Sarah Jane Shangraw, was a guiding star for me through the publishing process. Her encouragement and advice made this publishing process most rewarding, illuminating, and fun! My wife, Marianne provided not only moral support over the years, but reviewed and commented upon many versions of this work. Her sharp eye was always appreciated. My daughters, Stephanie and Hope, have been supportive in their own ways as well over the years. Many other scholars, friends, and associates—but unfortunately too numerous to mention individually—contributed in various ways to my work over the years, and I thank them all. I hope they know who they are. Of course, any errors of fact or interpretation are mine—mea culpa!

CHAPTER ONE

Introduction

Knowest thou the ordinances of heaven?
Canst thou set the dominion thereof in the earth?[1]

"How did Boulder happen?" asked Lewis Branscomb, creator of the Joint Institute of Laboratory Astrophysics in Boulder, in 1980 at the fortieth anniversary symposium of another Boulder scientific institution, the High Altitude Observatory. He went on to compare the city nestled in the foothills of the Colorado Rocky Mountains to another world center for scientific knowledge production. "Look at this wonderful town of Boulder—the Akademe-Gorodok of the USA," he added, referring to one of the USSR's science cities.

Implied in Branscomb's rhetorical question was an observation of immediate import to this dissertation. Unlike the centrally planned science cities of the former Soviet Union and the United States, Boulder's rise as a city of knowledge was sudden and in many ways unexpected. The amount of research that occurred there in such a short time period causes observers today to believe the developments were planned. During approximately the mid-1940s to the mid-1960s, a small city little known for scientific or intellectual accomplishment became a world-recognized center for scientific knowledge production. This book investigates the answer to Branscomb's rhetorical question, "How did Boulder happen?" In doing this, this research also helps answer an even broader question of science and technology studies—why do science activities often "clump" together?[2]

This work therefore explores the creation of a modern city of scientific knowledge production, a "city of the sun," to recall the words of Tommaso Campanella written in 1623.[3] Despite having little in 1945 to suggest its future as an international site for science, Boulder rose to prominence as a center of scientific knowledge production in less than two decades. This work argues that Boulder's development centered initially on a simple but compelling scientific question—how does the sun affect the earth? Without centralized planning, scientific entrepreneurs and various elements in the local community, including a state university, exploited opportunities to advance the science based on this question. A shifting combination of diverse sponsors presented these opportunities in the post–World War II and early Cold War era. Because of the nature by which Boulder rose to prominence in the world of international science, the investigation of the city affords a unique opportunity to study the relatively sudden co-development of a compelling scientific question, a university, and an American site of scientific knowledge production.

Factors removed from Cold War concerns, such as local politics and personalities, played significant roles in how Boulder developed into a city of scientific knowledge production. Local scientists, three University of Colorado presidents, and other supporters helped to create what noted French astrophysicist Jean Claude Pecker enthusiastically referred to as "AstroBoulder." For him, Boulder was a city of "many things" with respect to space and atmospheric science by the 1960s.[4]

These scientists and other supporters unintentionally transformed the city into "AstroBoulder" from what many perceived as a "scientific Siberia" in the late 1940s. They did this by making the relatively remote Boulder region a world-recognized place to investigate the sun–earth connection in its immense complexity.[5] While doing this, the scientists and others involved did not merely seize the numerous opportunities the Cold War–era context presented for their science and their city. As this research demonstrates, they often created, shaped, and presented those opportunities in the first place. To orient the reader to this discussion, this chapter briefly introduces the Boulder of today, sets the historiographic context for these studies, and outlines the structure of this narrative and analysis.

Boulder Today—A Center of Scientific Knowledge Production

Boulder is unique in the United States. This still small city of about 100,000 citizens in the foothills of the Rocky Mountains now claims to have one of the highest percentages of citizens with bachelors and advanced degrees of any city in the nation.[6] According to a survey in 2008, three-quarters of the population have at least a bachelor's degree, while 10% of the population holds doctorates.[7] *MarketWatch* also listed Boulder as the "smartest city" in the United States in 2014, outranking intellectually high-powered rival areas such as Ithaca (New York), San Jose (California), and Ann Arbor (Michigan).[8] The city's reputation as an intellectual powerhouse reflects the diversity of science research and technological development occurring in the city today. This reputation is the legacy of Boulder's development as a center of scientific knowledge production in the mid-twentieth century.

By the 1960s, Boulder achieved an international status as a center for environmental sciences. As early as the 1950s, much of the science done in Boulder related to the natural environment. Since the late 1960s and early 1970s, the city has housed many of the nation's leading environmental research organizations, including the National Geophysical Data Center and the Cooperative Institute for Research in Environmental Sciences. The data center serves as the nation's repository of comprehensive scientific information about the planet's environment from the interior of the earth to the sun. The latter institute, founded in 1967 and cosponsored by the National Oceanic and Atmospheric Administration (NOAA) and the University of Colorado, is one of the nation's premier organizations for the interdisciplinary study of the natural world. Especially with the U.S. government's creation of the Environmental Science Services Administration (ESSA) in 1965, Boulder established its reputation as world center of environmental research, as many of ESSA's component organizations resided in Boulder.[9]

Much of this interdisciplinary environmental science in Boulder built upon the city's foundation as a center for space and atmospheric sciences in the years following the Second World War. Boulder achieved a wide reputation as a place for this research. Despite the city's subsequent development into a place for many types of scientific and technological activities, it remains a world center for space and atmospheric science. Institutions such as the High Altitude Observatory (HAO) and the Central Radio Propagation Laboratory (CRPL) of the National Bureau of Standards (NBS) that formed the basis of the story of modern science in Boulder still exist, although in altered form. The observatory became part of the National Center for Atmospheric Research (NCAR) after the center came to Boulder in 1960. Employing over 100 PhD scientists, NCAR is today a leading center for atmospheric and climatic research.

CRPL evolved into NOAA's Space Weather Prediction Center. The center serves as the primary national civilian effort to monitor the sun and warn of solar activity that might have deleterious effects on the earth and human activities.[10] The prediction center, along with NOAA's other Boulder-based research labs, represents yet another indicator of Boulder's continuing prominence as a center of solar–geophysical studies.

The NBS laboratories that came to Boulder as part of the CRPL move in 1954 still exist as well. Greatly expanded to over 350 researchers and staff, they now form part of the U.S. National Institute of Standards and Technology (NIST).[11] The lab's work blossomed beyond its original focus on sun–earth research and radio propagation, thereby establishing Boulder as an important center of physical science research and its applications.

Other Boulder-based geophysical and astronomical research centers that began in the 1950s and early 1960s have expanded significantly in the last forty years. These include the NIST–University of Colorado Joint Institute for Laboratory Astrophysics (JILA) and the University of Colorado's Laboratory for Atmospheric and Space Physics (LASP), both major research centers in astrophysical and planetary science.[12] The joint institute currently boasts two Nobel laureates on its faculty, indicating its cutting-edge research in atomic physics. Costing in total hundreds of millions of dollars, LASP experiments currently monitor many aspects of the earth's atmosphere, the sun, and the outer planets.[13]

The university's original Department of Astro-Geophysics, despite having several name changes over the decades, continues to this day as a world center for astrophysical, solar–terrestrial, and planetary studies.[14] The continuity of these Boulder scientific organizations created in the 1950s underscores their importance to U.S. and international space and atmospheric science.

But the city today is much more than a center for scientific research alone. Building on the foundation of science institutions established in the 1940s and 1950s, "high tech" industries began migrating to the Boulder region in the 1960s. International Business Machines (IBM) set up a major plant near Boulder in 1965. A spin-off in the 1950s of the University of Colorado's Upper Atmosphere Laboratory, Ball Brothers Space Technology, prospered in the 1960s and 1970s with the advent of space-based astronomy and solar physics.[15] Beginning with operations such as Ball and IBM, Boulder and the surrounding county became a major and growing location for light high-tech industry and software development in the 1970s and after.

Boulder and U.S. Regions of Scientific Knowledge Production

The foundation of Boulder's development as a city of knowledge resulted from scientific endeavors that Harvard-trained solar astronomer Walter

Orr Roberts helped to create in the city in the late 1940s and 1950s. Roberts played a pivotal role in the story of Boulder's development as a modern city of science. Although he did not know it at the time, as Roberts drove west in the summer of 1940 on the way to Colorado, he began a process that would help create a modern city of scientific knowledge production.

Roberts brought the intellectual foundation for much of the city's post-1945 science activity—a better understanding of the physical nature of the sun–earth connection. The explanation of the city's development into a site for modern science, like the physics of the sun–earth connection, is complex, multilayered, and varies over time.

In the 1950s, Boulder became a city of knowledge—defined in part as a location that became one of the "engines of scientific production."[16] But what occurred in Boulder fundamentally differed from events in other areas of scientific and technological development in this era. What makes the Boulder story unique and informative rests in the unexpected nature of Boulder's rapid rise to the very top rung of U.S. centers of scientific research and associated activity in the early Cold War era.

Unlike the Massachusetts Route 128 corridor or California's Silicon Valley, Boulder possessed few of what Markusen and others refer to as the traits or indicators necessary for a high-tech region or a center of scientific knowledge. These traits include a local scientific labor force, a research university, resident companies, and an airport hub.[17] In Boulder, the scientific quest came first. Many of these growth indicators followed the science, and in some cases arose from the investigations of the sun–earth connection in Boulder initiated by Walter Roberts and others.

In this manner, the Boulder story stands in contrast to other sites of knowledge production of the time, such as the Cambridge–Boston area, Palo Alto, and Washington, D.C. These centers had many, if not all, of these growth traits. Even with its proximity to Denver about 40 miles away, Boulder paled in comparison to these other regions as a potential center of international science. This dearth of growth predictors, combined with the mountain city's distance from existing science centers in the northeastern United States, makes Boulder's subsequent development all the more intriguing.

In an attempt to explain the rise of scientific cities and regions, scholars often focus on one of the aforementioned growth traits or similar indicators. Authors Margaret O'Mara, Rebecca Lowen, Stuart Leslie, and Paul Ceruzzi argued to varying degrees that a prominent research university and associated scientific entrepreneurs provided the basis for the rise of these modern regions of knowledge.[18] By investigating areas that have preexisting major universities in the city or nearby, the historiography to date misses an important phenomenon. That is, how did a city of knowledge develop that did not have a well-established major research university in close proximity? It is precisely the absence of such an institution that makes Boulder's development instructive to study. As demonstrated in this research, the University of Colorado—a middling state university in the 1940s with only a very modest scientific research component at best—co-developed with Boulder to become a prominent site in the U.S. scientific and educational establishments.

Much of the pre- and post-WWII debate in U.S. science policy hinged on how best to fund American science. Some government policy-makers and other sponsors of science wanted to foster the growth of new centers of learning away from the eastern scientific establishment located in intellectual centers such as Cambridge, New York, and Washington, D.C. Others thought it crucially important to bolster these established centers, following the exhortation of the Rockefeller Foundation's Wickliffe Rose in the 1930s to "make the peaks higher."[19] To employ Rose's metaphor, Boulder lacked a noticeable scientific or academic "peak" when Roberts arrived in the mid-1940s. Rose did not seem much concerned with the creation of new centers of science. Contrary to Rose's thinking, this dissertation is concerned primarily with the creation of these scientific peaks in the first place.

In explaining the development of the regions around MIT and Stanford, researchers have argued that the role of these regions as key components of the burgeoning U.S. military–industrial complex played an integral part in their further development as scientific and high-tech centers. Leslie discusses the importance of a "golden triangle"—interrelated activities of the military, research universities, and high-technology industries—in creating these centers of science and technology.[20] Boulder, in contrast, housed at best only one side of this "golden triangle" in the postwar years—relatively modest levels of military funding for sun–earth research. By the late 1950s, much of this military funding gave way to the civilian funding of science. Even before this occurred, unlike in other regions, a significant source of Roberts's funding came from private donors.

The U.S. military through the mid-1950s was a major patron of Boulder's sun–earth connection research. As David DeVorkin, Ronald Doel, and others have described in detail, the Department of Defense played a crucial role in advancing space science and related research from 1945 to the mid-1950s, especially by encouraging the use of new tools such as the rocket (mostly in the form of V-2s from Nazi Germany).[21]

In contrast to much of what transpired in other science regions of the time, the then-modest scientific work in Boulder produced generic knowledge about the earth and its environment. Little, if any, scientific knowledge generated in Boulder sun–earth science entered classified realms after World War II. Much of this knowledge was not uniquely useful to the military— it was inherently "dual purpose," to use a term in use today. The military wanted to generate comprehensive knowledge about the physical world. This knowledge was not necessarily required for any individual weapon system. The Department of Defense sought this generalized knowledge simply because post-WWII U.S. military personnel and weapon systems operated throughout the natural environment—land, sea, air, and space.

The military perhaps had another (almost altruistic) reason to fund the earth sciences as well. Vannevar Bush, an architect of postwar U.S. science policy and then elder statesman of U.S. physical sciences, reflected on the role of military funding in the post-WWII era to the early 1960s. "The military not only support military research, they also support a lot of research which has very little to do with military objectives indeed," the noted policymaker opined. This largess for science occurred, Bush argued, because the interaction with the U.S. scientific community put the military "in touch with a large group of civilians." Therefore, Bush added, "we're getting along as well as we would have" if only civilian scientists determined U.S. science policy.[22] The knowledge generated by military-funded science was also useful to the broader society, such as in assisting civilian radio propagation, for example. Military authorities, from their own experience and association with scientists, understood the usefulness of this information to nonmilitary segments of society. As a result, military authorities did not find it necessary to classify most of this knowledge about the natural world.

With scientific results and data available to all researchers, Boulder science bears little resemblance to an intellectual "archipelago" that Michael Dennis argued existed in other parts of the U.S. science complex during the Cold War.[23] Boulder became a site primarily for widely available information about the sun–earth environment, even though the research remained chiefly funded by the military for approximately a decade.[24] Walter Roberts and others in the city had a deep commitment to what Robert Merton called "communalism" as a core value of scientists.[25] Boulder researchers—notably Roberts—chose not to do classified research, for they wished to communicate freely with other scientists working on similar research. As a result, atmospheric and space science in Boulder never had significant classified aspects. This allowed a free and open communication among researchers, not only locally, but nationally and internationally as well. As explained subsequently, this openness allowing free communication of science is a key aspect of the Boulder story.

As a result, the scientific activities of the city do not fit well into a traditional "military–industrial complex" analysis or description.[26] As a result of the military funded, but inherently dual-purpose, sun–earth connection work, Boulder found itself ideally suited to benefit in the post-*Sputnik* era. In this period, much funding for science transitioned from military to civilian sources, such as the National Science Foundation (NSF) and the National Aeronautics and Space Administration (NASA). As Boulder science was not science that only the military wanted done, the city was well poised to take advantage of the funding opportunities these new agencies offered.

It is important to note also that central planning, military or otherwise, did not factor into the Boulder story. Although superficially similar to cities such as Los Alamos in the United States or the centrally planned "science cities" of the Soviet Union, such as Branscomb's Akademe-Gorodok, Boulder developed in a fundamentally different way.[27] The process was completely ad hoc. At no time did anyone, anywhere have a detailed plan as to how Boulder should develop. The city's development was therefore a product of exploited opportunity by those who recognized it when it arose.

Boulder's development also differed from another front-range city in Colorado—Colorado Springs. Although both cities lacked central planning in their rise to prominence, the latter became a major site for U.S. military space activities. As such, Colorado Springs formed part of the military–industrial complex of the time—the "Gunbelt" of the United States—because of these militarized aspects.[28] Boulder, in comparison, established itself as a "scientific–academic" complex. As such, it represented a very different phenomenon in regional development from true military–academic complexes.

Robert Rosenberg, in his book *Boulder, Colorado: Development of a Local Scientific Community: 1939–1960*, describes in broad outline the process by which Boulder evolved into a city of science. It discusses the role of solar–terrestrial physics as the "research wedge" that allowed the development of science in Boulder. He also suggests an "accretion" model of the process of Boulder's development whereby layers of activity build upon each other.[29] This model, based on solar–terrestrial physics as the glue of the accretion, omits a detailed explanation of why new scientific entities arose there and why they "stuck." The city's development as a science center was horizontal, not vertical or hierarchical, so organizations did not simply build on one another. It might be tempting to see Roberts's High Altitude Observatory as a core or nucleus of science in Boulder, but Roberts was not an "empire builder" in the usual sense of the concept. To the contrary, he much valued smaller scientific organizations over larger ones.

The alternate idea proposed here echoes AnnaLee Saxenian and other's observations of Silicon Valley and similar technological regions.[30] A social dynamic model—based on the sun–earth connection as a means of diverse and open scientific interaction—has the robustness to explain the entirety of the scientific developments in Boulder, including the crucial roles played by the university, the public, or the local business community in developing Boulder as a science city. All are fundamental to the story of how science arrived in Boulder.

Central to Saxenian's analysis are ideas that serve to best explain much of the dynamic in Boulder's development as a site for science, especially in its early phase in the post-WWII era. This includes the important role of local culture of the free exchange of knowledge—openness—in explaining Silicon Valley's success in the rise of semiconductor technology and industry in the 1980s. Coupled to this is an explanation of the associated failure of Route 128 near Boston to successfully adapt initially to these new technologies—there was much less open exchange of knowledge in this region. Even though both regions had military funding available, Silicon Valley, as a result of its culture of openness, dominated this and related industries.

In the Silicon Valley case, a regional industrial system promoted "collective learning and flexible adjustment among specialist producers of a complex of related technologies."[31] Although not entirely analogous, for in Boulder the "product" consisted of knowledge and not material devices or technologies, Saxenian's insights nevertheless do apply to Boulder's development as a science center. A relatively small group of researchers and their associated organizations collaborated in an adaptive and open manner to advance the diverse aspects of the sun–earth science that motivated them. In doing this they created a "Marshallian district," a place where skilled workers with specialized knowledge congregate and produce an industrial center of some type.[32] Implied in this concept is a feature that also existed in Boulder—horizontal, versus vertical, integration of activities, whereby research sites could coordinate and interact without overarching guidance from a central authority.[33] With horizontal integration, the emphasis is the interdisciplinary creativity arising from the cooperation of disparate but geographically related activities.[34] The twist in Boulder was that the district centered on science, not industry. In doing this, beginning with Walter Roberts, they set the foundations for the city's later development as an environmental science and high-tech center as well.

Roberts established himself at the leading entrepreneur of science in Boulder after he relocated there in 1946. He sought sponsorship from a wide spectrum of interests to fund not only his own research, but other sun–earth science in Boulder as well. As we will see, Roberts (and Boulder) benefited from the initial military-sponsored largess based on the armed services' desire to better understand the natural environment in which they operated.

In a time of debate over the best way to fund science, Roberts "split the difference" in these positions by soliciting both government and private funds. Although the U.S. military provided the majority of his funds through the 1950s, Roberts successfully obtained private funding from a host of entities to augment the military support. Roberts showed a strong preference to adhere, at least in part, to the earlier tradition of the private funding of U.S. astronomy. Patrick McCray delineated the sharp contrast between the more traditional views of astronomers like Jesse Greenstein, who depended on private support, and others such as Leo Goldberg, who advocated a much larger role for government funding.[35] After *Sputnik* (in October 1957), Roberts easily made the transition to accepting full government support for his and other associated sun–earth research in Boulder, and his private fundraising efforts disappeared in the wake of increased U.S. direct civilian funding of pure science after 1957. His lack of any classified military research made this transition straightforward and seamless.

Many American scientific institutions benefited from increased government funding in this post-*Sputnik* era. As Roger Geiger demonstrates, U.S. universities figured prominently among those that gained from this new government munificence.[36] The administration of the University of Colorado, in part inspired by Roberts and others in Boulder, understood the importance of becoming a center for sun–earth science. The university then took advantage of the new funding possibilities for this type of science. As University of Colorado president Quigg Newton often said referring to these times of funding largess, "There is a tide in the affairs of men. / Which, taken at the flood, leads on to fortune."[37] This was a tide he and others rode happily. Because of the particular path of early science in Boulder, a strong military link never existed at the University of Colorado—unlike MIT or Stanford, for example. Science at the school developed mostly in the wake of *Sputnik*, not before it.

Roberts and the university existed as part of a local community. The continued support of a community proved crucial for the development of Boulder as a city of science. As frequently asserted in regional studies relating to the development of scientific and technological innovation, local "boosters" often played critical roles in that development.[38] These boosters in

Boulder included elected officials, business people, community organizers, and everyday citizens. Some of them gained directly or indirectly by bringing research science to Boulder, others did not. The explanation of events in Boulder remains incomplete without an accounting of community booster activity as encompassed in this study.

All of these entrepreneurs and supporters of science in Boulder—Roberts and other researchers, the University of Colorado officials, and local boosters—exploited the compelling idea of the sun–earth connection in this beginning era of modern Boulder science. By providing the supporting rationale for diverse sponsors to support research in the city, sun–earth research helped to establish the foundations of today's Boulder.

Boulder and the Sun–Earth Connection

How did Boulder develop as a sun–earth science city in the postwar era?

An important piece of nomenclature needs definition at this point. In the early twentieth century, the terms "solar–terrestrial physics" and "solar–terrestrial relationships" assumed currency among practitioners of these sciences. These terms described the suite of studies that involved understanding the solar–terrestrial environment and associated effects on earth. Roberts and Harvard astronomer Donald Menzel, his academic mentor, eventually introduced in the 1950s the label "astrogeophysics" for this brand of research.[39] In the last few decades, the more descriptive and concise term "sun–earth connection" has often replaced the older (and somewhat more ambiguous) terms in discussions.[40]

There is little historiography, relatively speaking, about sun–earth connection science, and even less exists on Boulder's important role in these studies. Many refer to the 1950s and 1960s as the beginning of the "Space Age."[41] Historians, with a few notable exceptions, including David DeVorkin in *Science with a Vengeance*, Karl Hufbauer in *Exploring the Sun* and Ron Doel in *Solar System Astronomy in America*, have paid relatively little attention to the history of space and atmospheric sciences in this formative era. These works highlight the importance of military funding in the period, and Hufbauer demonstrates the role of newer scientific funding entities in developing space science after *Sputnik*. In addition, except for some discussion in Hufbauer's book, no historical work focuses on the importance of the sun–earth connection (defined subsequently) as a compelling scientific question. Some works do further discuss the development of geophysics and related disciplines in the post-WWII era. The October 2003 issue of *Social*

Studies of Science, dedicated to the earth sciences, included a number of useful articles relating to the space and atmospheric sciences.[42] In particular, Doel's "Constituting the Postwar Earth Sciences" provides context and overview for the current research by showing the military's role in fostering the earth sciences to varying degrees.[43] Doel's essay relating to discipline formation, "Earth Sciences and Geophysics," in *Companion to Science in the Twentieth Century* also helps us to better frame events in Boulder.[44] This article, a history of environmental sciences in the twentieth century, traces how earth sciences in this era set the foundation for "environmental science" soon after. Michael Dennis's essay in the *Social Studies of Science* volume on the Cold War and earth sciences provides insight into the nature of earth science disciplines and their various funders in this era.[45]

The inherently interdisciplinary aspect of these multifaceted investigations of the earth's complex physical environment was a determining factor in many of the events in Boulder's rise as a center for knowledge production. Sun–earth studies included not only broad swaths of astronomy and physics, but chemistry, atmospheric science, radio propagation research, hydrology, fluid dynamics, and many others. Sun–earth connection research served as more than a "research wedge" for bringing science to Boulder.[46] The diversity of the studies required for productive sun–earth research provided a robust and wide foundation to Boulder's subsequent development as a center for science in the 1950s—an unintended, but important, consequence of Boulder becoming a home for sun–earth science in the late 1940s and early 1950s.

There exist few, if any, archive-based detailed studies of the development of Boulder as a world center of scientific knowledge production in the foundational years 1945 to 1965. *Quest* magazine published David Spires's scholarly account of the early Boulder story, centering on Walter Roberts's role as a catalyst there.[47] It elucidates Walter Roberts's crucial role in developing "Boulder's Aerospace Community." The article focuses on the pre-*Sputnik* era, and as a result, the complex story of how Boulder became the site for NCAR in the late 1950s and early 1960s is not covered.

Others have focused research on more limited aspects of the Boulder story. In 1970, Elizabeth Hallgren authored an organizational history of the University Corporation of Atmospheric Research (UCAR), including its associated High Altitude Observatory and NCAR.[48] NSF historian George Mazuzan wrote a comprehensive account about the role of UCAR in the creation of NCAR and Roberts's role in this process.[49] As the work limited its treatment to the creation of NCAR, Mazuzan does not deal in any detail with the important issue of NCAR site selection and Roberts's heavy role in this.

Structure of this Book

To facilitate an understanding of the events that led to Boulder's establishment as center of space and atmospheric science knowledge, we employ a chronological narrative. Walter Orr Roberts, almost unknown today in the city, was a pivotal figure in creating modern sun–earth research and establishing science in Boulder. He maintained, beginning in the mid-1940s, an extensive collection of personal papers that now resides at the University of Colorado. In addition, UCAR maintains an extensive archive of not only Roberts's correspondence, but of HAO and NCAR documents as well. These collections form the foundation of this research.

Because of his importance in the Boulder story, the first chapter briefly presents Roberts's early scientific training, explains how he got to the mountains of Colorado in the immediate WWII years, and discusses his scientific activities during the conflict. This chapter demonstrates how this scientific activity in the Rocky Mountains started as Harvard's attempt to establish scientific outposts in the American West by Roberts's mentor, astronomer Donald Menzel—his "western observatories" as Menzel later called them.[50]

The second chapter relates the story of Roberts shifting activities from Climax Station high in the Rocky Mountains to Boulder, situated in the foothills and about forty miles northwest of Denver. His initial move occurred in large part to support a new scientific entity, the High Altitude Observatory, a joint effort by Harvard and the University of Colorado. The University of Colorado in Boulder strongly encouraged this relocation, and the institution plays an important role throughout this narrative. This chapter also illustrates how a competitor to the Boulder operation soon arose—the Sacramento Peak Observatory in New Mexico. Roberts's intervention helped to keep HAO in Boulder, setting the stage for the future development of the city as a site for U.S. science.

Chapter 3 describes how the first major component of national science arrives in Boulder, the Central Radio Propagation Laboratory of the National Bureau of Standards. Scholars only have partially understood the involvement of Colorado's elected representatives in bringing the radio lab to Boulder, but here we attempt to clarify this aspect of the story. In addition, the chapter analyzes the vital role Boulder residents had in these events. The local community provided the impetus for more scientific organizations coming to their city as the 1950s progressed.

Chapter 4 analyzes developments in Boulder as a city of science knowledge generation as the local, national, and international contexts of space and atmospheric science changed. This chapter includes a detailed study of

Roberts's efforts at institution building and related fundraising activities. It also addresses the personal split between Roberts and his mentor Menzel. The split resulted in the separation of the High Altitude Observatory from official ties to Harvard—Roberts no longer needed approval from anyone at Harvard to engage in new scientific initiatives. Occurring in 1955, this separation placed Roberts in a better position to benefit from the massive changes in U.S. science funding policy after *Sputnik*.

Chapter 5 discusses how Boulder became an important front in what journalist Walter Sullivan called the "Assault on the Unknown."[51] The International Geophysical Year (IGY) of 1957–58 is perhaps the most significant event for earth and space science in the twentieth century—it directly and indirectly spurred many investigations relating to the sun–earth connection. This chapter also shows how Roberts and close colleague Alan Shapley, a succession of presidents of the University of Colorado, and others understood that Boulder would be a key place to contribute to this new phase of the exploration into the poorly understood sun–earth connection. They also understood that Boulder, in turn, could benefit greatly from participating in these new scientific endeavors as well.

The launch of *Sputnik* in October 1957 as part of the IGY campaign had a profound effect on the development of Boulder as a science city. The landscape of American science altered suddenly and significantly. Chapter 6 discusses how changes in national atmospheric science policy propelled Boulder—in large part due to Roberts's efforts—to the top levels of atmospheric science research in the nation. This chapter analyzes how the National Science Foundation's building spurt of new national centers of science led, in close collusion with atmospheric scientists, to the creation of the National Center for Atmospheric Research.

Chapter 7 presents in detail the dynamics by which Roberts not only became the first head of this important activity, but ensured that the NSF located NCAR in his home city. As happened with CRPL's move west a decade earlier, the citizens of Boulder directly contributed to bringing more science to Boulder. The resulting, complex story of the "Blue Line" debate on limits to growth in Boulder is an integral part of NCAR's move to the city in 1960. Boulder sun–earth connection research and related science blossomed in the early 1960s. Chapter 7 therefore also explores how in the mid-1960s Boulder became a world center for more expansive environmental research—a natural development arising from the city's existing sun–earth science activity. The narrative ends in 1965, when the U.S. government created the Environmental Science Services Administration (ESSA) to consolidate national efforts dealing with the environment. The creation of ESSA solidified Boulder's reputation as a national center for scientific knowledge production—many of ESSA's component organizations were in Boulder. ESSA's creation also set the stage for Boulder's further development as a center for environmental sciences and concerns.

The conclusions highlight how the changing sponsor and funding patterns for American science in the time span under investigation affected developments in Boulder. We also discuss how the context of science policy in this era led to the development of an unlikely candidate into a modern city of scientific knowledge production. Was it "Uncle Sam" or entrepreneurial local citizens who were ultimately responsible for the city's rise as a science center? As Ceruzzi wrote in answering a similar question about Virginia's Tysons Corner, "it was both," and in Boulder's case, then some.[52]

CHAPTER TWO

Sun–Earth Science Arrives in Colorado

Solar Research and the Sun–Earth Connection before 1945

The notion that the sun could affect many processes here on earth, and not just simply be a constant source of heat and light for the planet, manifests a long history. Even the famed astronomer William Herschel argued around 1800 that solar processes indicated by varying sunspot activity could affect earth weather and related grain production.[1] In the nineteenth century, some scientists came to realize other phenomena associated with the earth, such as geomagnetic storms and the northern lights, might similarly have a connection to solar flares. By 1900, British physicist Sir Norman Lockyer could exclaim that "the eleven year period is not one to be neglected" in thinking about solar changes and effects on our planet, referring to the eleven-year sunspot cycle.[2] Notably, Samuel P. Langley and Charles G. Abbot of the Smithsonian began a decades-long study of the relationship between solar luminosity and the earth's weather in the later part of the century.[3]

By the 1930s solar astronomy matured as a scientific discipline by importing observational methods and theoretical frameworks from physics.[4] Still, the study of the sun did not form part of mainstream astronomy in the United States.[5] To astronomer Walter Roberts, the solar astronomy in this era needed both "a demand" and "a constituency" for further development.[6] Roberts, learning from his mentor, astronomer Donald Menzel at Harvard, eventually found that questions relating to the sun–earth connection could provide the stimulus to generate the demand and constituency needed for solar research to progress.

The creation of a center of scientific and technological production often results from the work of a single individual such as Walter Orr Roberts. Technical leaders associated with scientific centers include electrical engineer Frederick Terman at Stanford, referred to occasionally as "The Father of Silicon Valley,"[7] and Charles Stark Draper, who played a similar role at MIT and the area around Boston.[8] Other regional developers were nontechnical, business-oriented "promoters," such as Joe Reich of Colorado Springs.[9] All these individuals knew how to exploit both the historical circumstances and institutional contexts in which they operated.[10]

Such was the case with Roberts and Boulder's development. Roberts brought sun–earth connection science to Colorado in 1940 and eventually served as a major promoter of science in Boulder. In order to understand this development, it is important to examine his influences and interests, why he relocated from Harvard University to the Rocky Mountains, who funded sun–earth research at the time, and the nature of sun–earth connection studies in the 1930s and early 1940s. This chapter details the context of an up-and-coming solar astronomer and the state of one small aspect of solar science in the 1930s—the study of the sun's corona. Even as Roberts trekked west from Cambridge in 1940, he pondered questions relating to the sun–earth connection and how his studies might benefit humanity in addition to advancing solar physics as a scientific endeavor.

The Sun–Earth Connection and Dust Bowl America

In the summer of 1940, the Midwestern landscape that Harvard astronomy graduate student Walter Orr Roberts drove across to the Rocky Mountains had suffered greatly from years of severe drought. An extreme heat wave had produced the infamous, Depression-era "Dust Bowl."[11] Entire populations relocated to other parts of the United States to escape its effects. Even though the region's climate had improved somewhat by the late 1930s, evidence of the drought years remained into the 1940s.

The effect of this extreme weather on the landscape impressed itself upon the twenty-five-year-old Roberts, the son of a Massachusetts farmer. Roberts, already a trained solar physicist, asked himself a simple question: could changes on the sun ninety-three million miles away have generated these severe droughts of the 1930s? The trip, Roberts later claimed, greatly stimulated prior interests. "I've always had," he later told an interviewer, "a very strong orientation towards the practical uses of science, and so I sort of concluded that maybe the most important thing that could come out of solar physics research would be to understand and maybe predict the effects of solar activity on the weather."[12]

Roberts's simple question about the relationship between the sun and weather yielded complex answers, resting on the diverse and disparate physical mechanisms that coupled solar and terrestrial processes—the sun–earth connection. Much of his subsequent professional career centered on his attempts to address the sun–earth connection from various scientific perspectives and disciplines. Much of his career was also about how to best gather funds for these arcane studies.

The Education of a Solar Astronomer

Born on August 20, 1915, in West Bridgewater, Massachusetts, Roberts began his life and scientific career close to the Atlantic coast, just south of Boston. As a somewhat sickly, absent-minded youth, following in his father's footsteps as a farmer was not an option for Roberts. He developed, however, an early interest in astronomy, inspired by a local amateur astronomer and engineer, Lincoln K. Davis. Davis served as a big brother and mentor to the curious youngster. "He was a magnificent man," recalled Roberts about Davis. "I used to go to Linc Davis's house and work and play at his machine shop, and that's how I got interested in building things."[13] Davis also brought Roberts to large regional gatherings of amateur astronomers, further developing his interest in astronomical activities. These early influences shaped Roberts's choice of scientific pursuits, and he carried these interests through college and graduate school.

In 1932 Roberts enrolled in Amherst College. While there, he took all offered astronomy courses and worked at the college observatory as an assistant. Eventually, he took enough courses to not only formally major in physics and have a minor in math, but also to have majored in astronomy and chemistry as well. Roberts decided to study physical chemistry, and he claimed later that his goal during this period was to become president of Kodak.[14] In the Depression era there were few jobs in astronomy, and this seemed the most prudent career path to Roberts.

Upon graduation, Roberts began what he thought would be a long career at Kodak. However, Roberts's tenure as a full-time employee did not last long. He impressed administrators with his astronomical interests, so the company offered the young scientist a graduate fellowship. Roberts recalled that "Kodak had an incredibly far-seeing" educational support program, a program that could last "almost indefinitely as long as you made good progress."[15] The Kodak fellowship, along with an Amherst fellowship sponsored by his astronomy professor, Warren Kimball Green, enabled Roberts to begin his graduate studies.

Despite an original desire to attend MIT, Roberts altered his plan after discussing his plans for a career at Kodak with some of his Amherst professors. They advised him to study with Harvard's Otto Oldenberg, a noted experimental physical chemist. Roberts enrolled at Harvard in the fall of 1938. Given his work at Kodak, he initially chose to study not astronomy, but physical chemistry. Roberts hoped that he could still do astronomy research via Kodak's work on improving astronomical film.

Although he enjoyed studying with Oldenberg, Roberts had difficulty in a physics course taught by Wendell Furry. "I couldn't hack it. It was too tough for me," Roberts reflected. Despite the excellent education Roberts thought he received at Amherst, his training in advanced mathematics was deficient. Dropping this course so early in his graduate career altered his professional and personal life forever—it put Roberts in contact with Harvard astrophysicist Donald Howard Menzel, who became the dominant influence in his early professional career.[16]

"Providence" sent him to me, Menzel said of Roberts at a dinner in 1970.[17] It turns out, however, that Oldenberg sent Roberts to Menzel. "Oldenberg suggested it," said Roberts, referring to Menzel's spectroscopy course as an alternate to the dropped physics course. The only course that fit his schedule, Menzel's offering also worked well with Roberts's long-term goal of being a physical chemist.[18] The course excited Roberts and convinced him to forsake his career at Kodak to pursue his astronomical interests.

The academic change of plans brought Roberts into contact with another mentor who helped shape the contours of his professional life: influential Harvard astronomy department head and Harvard College Observatory director, Harlow Shapley. This began a fifteen-year personal friendship and professional relationship with the much older astronomer, famed for his discovery of the shape of the Milky Way.[19]

Donald Menzel had a profound effect on Roberts's life, and the two scientists would have an intense, complex, and close professional and personal relationship until the mid-1950s. Menzel's astronomical interests are essential to understanding subsequent events in Boulder. It is important to note that Roberts's mentor grew up in Leadville, Colorado—at 10,200 feet, the small mining town was the highest city in the United States.

Menzel attended the University of Denver and graduated with an AB (bachelor's) degree in chemistry in only three years. After earning an MA degree in 1921, Menzel moved to Princeton to pursue his doctoral studies. Once at Princeton, Menzel fell under the sway of Henry Norris Russell. Russell was a major force in establishing modern astronomy in the United States. He did this in part by creating the new discipline of theoretical stellar astrophysics. Menzel quickly mastered both the most advanced physical and mathematical concepts of the new astronomy.[20]

Because of his astrophysical work, Menzel's long relationship with Harlow Shapley and Harvard began. Menzel spent his third year in Cambridge studying stellar spectral line intensities in an effort to relate a temperature scale to the Harvard classification scheme of stellar spectra.[21] After finishing his doctorate in 1924, Menzel performed groundbreaking work on understanding the structure and composition of the sun's atmosphere at California's Lick Observatory. Knowing of Menzel's pioneering work, Shapley offered him a position at Harvard. The Harvard College Observatory (HCO) stood at the forefront of astrophysics and therefore "was everything Lick was not."[22] Menzel accepted the position.

After arriving in Cambridge, Menzel led large efforts to combine the theory and observation of gaseous nebulae in interstellar space. In doing so, he furthered Shapley's aims to make Harvard and the associated HCO a primary U.S. center for the rapidly developing discipline of theoretical astrophysics in the mid 1930s. In the "highly competitive" atmosphere of the Harvard astronomy program, Menzel established the careers of many of his talented students by incorporating them into the process of creating this new center for astrophysics, and included in this group of promising students was Walter Roberts.[23]

The Coronagraph and Sun–Earth Connection Enter Roberts's Life

The other dominant influence in Roberts's early career, Harlow Shapley, had a reputation for his eclectic interests, including ants, but a research interest in solar astronomy was not among them.[24] As director of the HCO from 1921 until 1952, he was a key figure in the development of modern American astronomy. His personality fit well into the "imperial" tradition of an extremely powerful director that dominated American astronomy of the era.[25]

Shapley understood that interdisciplinary approaches combining physics, chemistry, geophysics, and other disciplines with astronomy might serve to stimulate research across a broad front. This is a major reason why he permitted Fred L. Whipple's meteor study to proceed, a subject ostensibly removed from Shapley's stellar and galactic interests.[26] He fostered inherently interdisciplinary sun–earth research by similarly permitting Menzel to pursue his interests. This interdisciplinary approach proved vital to stimulating not only large sectors of American astronomy, but the young Roberts as well.

As a result of his ambitions for Harvard astronomy, Shapley needed help in developing a solar studies program. Because of his position, Shapley wielded much influence on the discipline by shaping local research programs and controlling the main components of astronomy's lifeblood: observing time and funding.[27] Shapley's own research concerned galactic structure, and he showed little interest in funding solar work. Shapley encouraged Menzel to undertake these solar investigations, but Menzel had to develop his own source of funds. Yet by hiring Menzel, Shapley established the foundation for a viable and productive program in modern solar astrophysics at Harvard.

Roberts had entered serendipitously into Menzel's plan to expand solar research at Harvard. Menzel wanted to do this by bringing to Harvard intense interest in solar eclipses. Besides the awe generated by the phenomenon, an important scientific feature of eclipses is that they reveal, briefly, the outer atmosphere of the sun. Astronomers called this solar feature the "corona," because it appears similar to a crown around the sun. Astronomers long noted the importance of getting the short-lived (flash) spectra of the sun during a total eclipse. By analyzing the light from the corona revealed at these times, scientists could begin to understand the structure, composition, and temperature of the sun's atmosphere. Solar eclipses provided astronomers with important opportunities to better understand stellar atmospheres. Solar coronal research formed a fundamental part of Menzel's program at Harvard and became an intimate part of Roberts's academic life.

Unfortunately, a solar eclipse phenomenon lasts but for a very short period—seven or eight minutes at most—as the moon's shadow races across the surface of the earth. Astronomers puzzled in the late 1920s and early 1930s over the possibility of producing artificial eclipses at one's choosing. Could they somehow generate a solar eclipse at will?

The "yes" to the answer of artificially generated eclipses came from France. Bernard F. Lyot, an astronomer at the Meudon Observatory near Paris, built the first truly functioning instrument with which one could observe the sun's outer atmosphere at will. A device that simply blocked the light of the solar disk with an occulting disk somewhere in the light path of the telescope did not present a challenge to telescope builders. The real problem concerned residual light in the telescope's tube, and even minute amounts of stray light ruined the image of the corona.

Lyot overcame this problem of scattered solar light in the system by using a polarimeter that blocked unwanted light and thus succeeded in building the first successful "coronagraph."[28] Initially used on July 25, 1930, at Pic du Midi Observatory high in the Pyrenees to examine solar features in the solar atmosphere known as "prominences," Lyot's invention revolutionized solar studies.[29] By successfully tackling a problem that stumped many of the world's finest observational astronomers, Lyot achieved an international reputation.[30]

A few years later, Roberts commented that Lyot's device to artificially duplicate eclipses at any time was "one of the truly significant achievements of the entire history of astronomy."[31] The clever device advanced solar and sun–earth research and played a key role in Roberts's life from then on.

Menzel also attempted to build a similar coronal observing device in the early 1930s. Menzel remained unaware of the French coronagraph in the early 1930s, though precisely when and why he decided to build one remains uncertain.[32]

The first question Menzel had to address was not a scientific one. Rather, the issue for Menzel was more earth-bound—where to get funds to build the device? Fundraising, from both private and government sources, formed part of Menzel's life at this time. Autocratic Harlow Shapley insisted to Menzel that there would be no funding forthcoming for the coronagraph or other solar work. Even though Shapley saw merit in the solar research, he told Menzel that if he needed funds for the effort, he "would have to go out and raise them."[33] In this era of the "imperial" model of observatory directors who dominated American astronomy that together formed a "power elite," there was little recourse for Menzel and others astronomers wanting to undertake research projects of their own choosing.[34]

For Menzel, and other astronomers, securing these funds was a difficult task. Increasingly sophisticated instruments, such as the coronagraph, required commensurately more money to construct. American astronomy in the 1930s and before mostly consisted of individuals or small groups working on relatively small amounts of private funding (compared to post-WWII), interspersed with a few large-scale projects also supported by private donors.[35] The Palomar Observatory, with its 200-inch mirror then under construction, exemplified such a project. Although usually small by subsequent standards, these grants from individuals often determined the success or failure of these smaller projects.[36] This is not to say that there were few funds available for astronomical research, even in Depression-era America. Astronomy in the nation before WWII did fare relatively well compared with other sciences, garnering almost a third of the funds private endowments allocated in the period.[37] This did not, however, provide copious amounts of money for astronomers, as most funds supported highly competitive larger projects such as Palomar. Menzel therefore had to search far and wide for money.

Disparate sources eventually sponsored the Harvard coronagraph project in its formative years of the late 1930s and early 1940s. Observing how Menzel obtained this support was a crucial part of Roberts's graduate experience, for these fundraising and entrepreneurial skills were eventually crucial to his own activities as a scientist and scientific entrepreneur.[38] Roberts learned from among the best, for Menzel was, in the words of a long-time colleague, "a great salesman always."[39] This aspect of his education centered on Menzel's specific attempts to convince a diverse group of potential sponsors that there were practical, mundane benefits to understanding the sun. There was a "sun–earth connection," Menzel suggested, that should concern everyone. This connection, in various ways, affected every human activity.

Menzel, and later Roberts, chose to assume what one scholar called the "CEO" style of observatory directorship, a style that required in part great skill at fundraising. Some astronomers, such as Shapley's HCO predecessor, Edward Pickering, went so far as to employ a public relations expert to garner funds.[40] Shapley himself aggressively sought funds from the National Academy of Sciences for HCO work.[41] Menzel brought funding to a new level after discovering the possibility of exploiting the sun–earth connection as a fundraising device for his solar work. He did not appeal to altruistic ideals such as "knowledge as its own reward" to potential donors. Menzel's appeal, rather, centered on the mundane and practical uses of sun–earth connection knowledge.

Why would anyone in business or government find this Baconian appeal to the practicality of sun–earth research appealing? Because of the known and suspected links between the sun and earth, Menzel argued, scientists had to investigate the physical processes behind these links. By studying the sun–earth connection, scientists could better predict solar events and understand their effects on human activity in general (and more narrowly on the sponsor's interests).

At the time there existed ample reason to speculate on the practical benefits of sun–earth connection research. In the 1930s, scientists, particularly at the U.S. National Bureau of Standards, understood that there was a relationship between solar activity, such as the explosive phenomenon near the sun's surface called "flares," and problems with radio communications.[42]

Even well before the 1930s, scientists such as John Herschel in the mid-nineteenth century suspected a relationship between sunspots and compass disturbances during "magnetic storms"—a term coined earlier in the century by the naturalist Alexander von Humboldt.[43] The relationship between these phenomena so stimulated European scientists and naturalists that Edward Sabine in England organized a worldwide "magnetic crusade" (with Humboldt) beginning in the late 1830s to address the problem.[44] The first effort of its kind, the crusade prefigured subsequent International Polar Years and the International Geophysical Year of 1957–58.

Menzel needed a similar sun–earth link to which potential sponsors could relate, and that link was the weather. By understanding phenomena such as the nature of coronal disturbances via the coronagraph, he argued to possible donors, scientists could ultimately better predict the weather and its effects. Even though it was strenuously advanced at the time by a few noted scientists, including the Smithsonian's Charles G. Abbot, this relationship was controversial among both solar astronomers and meteorologists, and it remains so to this day.[45] The claim still enjoyed some currency in the 1920s and 1930s because of Abbot's strong advocacy. Much of the controversy then surrounding Abbot centered not on the possibility of a sun–weather connection itself, but on his attempts to find a link in the variations of solar luminosity with atmospheric phenomena.[46]

Irrespective of the state of the controversy about the nature of the possible sun–weather connection, Menzel and Roberts appreciated the question on its scientific merits. They also understood its potential value as a fundraising tool. Easily understood by even the scientifically unschooled, the sun–weather appeal for funding had the best chance of working with a very wide range of potential sponsors. This appeal translated arcane knowledge about the sun into ideas non-scientists could easily grasp as potentially useful to their own interests and to society at large.

Menzel's sun–weather connection approach proved a wise strategy for the most part. It occasionally worked. The most notable success was with Depression-era Secretary of Commerce Henry Wallace. He decided to give Menzel the small grant of about $5,000 in 1935 that helped start the coronagraph project after Shapley declined funding.[47] Wallace was eager to help American farmers and was a trained scientist himself. He was also on friendly terms with Shapley and Harvard president James B. Conant and understood the potential payoff of better weather forecasting for American (and world) farmers.[48] The Wallace funding ran out in 1939.[49] The sun–earth connection approach, however, proved fruitful for Roberts many times in his subsequent career.

Armed with Wallace's funding, Menzel designed his first system—the "coronavisor." Menzel hoped to use electronics to solve the problem that Lyot effectively dealt with by optical design. Menzel's system did not work well, perhaps because he never took it to a high altitude as Lyot did with the Pic du Midi coronagraph.[50] Menzel did not directly correspond with Lyot for help until February 1938. This contact occurred only after Menzel abandoned his coronavisor system when it proved unworkable, and then had trouble duplicating Lyot's device.[51]

The serious problems he encountered in coronagraph construction, along with reduced funding for non-Harvard help, may also explain why Menzel brought in enthusiastic graduate student Walter Roberts to work with him on the corona project in the late 1930s. As reported in a memoir by Roberts's son, David, the elder Roberts wrote in his private journal that he kept during the first year at Harvard that "I must work harder," with "must" being triple underlined.[52] Roberts delighted in building anything mechanical, so working on the coronagraph seemed a natural path for him. At the time, however, he apparently harbored hidden doubts about his abilities and chance for success in an astronomical career. Roberts began the work and later claimed that he "essentially built the coronagraph with my own hands, from the pile of parts, many of which didn't work."[53] He inherited this "pile" from a lab assistant Menzel let go because of reduced funding from the Department of Commerce. A few months after Roberts joined the effort there was a completed coronagraph at HCO's Oak Ridge site, about 30 miles west of Cambridge. The coronagraph did not work well, as there were still optical and structural problems with the device.

Menzel Points to Colorado

Menzel eventually understood that he needed the clearest skies to obtain the best results from his new instrument.[54] The near-sea-level HCO Oak Ridge station close to Boston was far from an ideal location. Following Lyot's example of stationing his device at Pic du Midi in the Pyrenees, Menzel and Roberts sought similarly clear skies. The choices of location narrowed quickly.

The mountains of Colorado, Menzel's home state and a region he knew intimately, served as a promising starting place. In the late 1930s, few places existed in the United States where one could find the roads and infrastructure required to support an observatory at high altitudes. "Climax was one of them," as Roberts declared later.[55] He referred to the small mining operation of the Climax Molybdenum Company (CMC), not far from Menzel's boyhood home of Leadville. For Menzel, ease of transportation was not the only issue, for he not only had his eyes on the clarity of the sky, but also on potential patrons—mostly businessmen and research foundations—as well.

The Cambridge-based astronomer headed west. Armed with a commission to act as a Harvard representative to obtain scholarship funding from successful alumni in the region, Menzel eagerly discussed both potential sites and financial support for the HCO coronagraph station. In a clear ploy to solicit local funding, he was not averse to gaining publicity for the potential site—a headline of the May 28, 1939, *Denver Post* declared "Colorado May Get Big Solar Observatory." The accompanying article highlighted that the region was the first to get consideration for this astronomical prize.[56] Echoing what had already become the standard Menzel line, the paper invoked the sun–earth connection by reporting that the observatory might generate new understanding with respect to long-range weather forecasting and other natural phenomena.

Menzel's efforts in Denver failed. Another potential location for the new observatory appeared around Pikes Peak, about fifty miles south.[57] Menzel visited nearby Colorado Springs and approached wealthy Harvard alumnus Spencer Penrose, an individual well known for his desire to develop the city.[58] One account of this visit relates that Penrose, apparently a staunch Republican, refused to even see the astronomer about the possibility of supporting a local solar observatory, apparently put off by Harvard's decision to give an honorary degree to the New Deal Democrat Secretary of Commerce Wallace in 1935.[59] Another slightly different version of the story explains that Penrose recoiled from Menzel's request because Menzel, as instructed by officials at Harvard, also asked the Colorado businessman for a donation directly to the university in addition to help setting up a local observatory.[60] In any case, no support for the coronagraph operation emerged after the trip to Colorado Springs.

Shaking off Penrose's rebuff, Menzel decided that Front Range mountain tops did not have good observing conditions for the coronagraph site anyway. Blowing dust from the Great Plains and turbulent observing conditions along the Front Range made the region a poor place for solar observing.[61] Menzel therefore looked farther west toward the Colorado mountain passes. Fremont Pass, at 11,000 feet in elevation and on the road from Leadville to

Aspen, provided a confluence of good observing conditions, passable roads, and a very willing sponsor: Max Schott, president of the Climax Molybdenum Company.

Menzel hit pay dirt with the president of the mining company. Schott's reasons for sponsoring solar research remain obscure, but perhaps Menzel's enthusiastic and well-argued appeal moved him. Schott immediately and enthusiastically decided to support in every way possible the building of the solar observatory at Fremont Pass, close to the CMC mining operation at Climax. Schott did not just give Menzel land for the observatory; he offered to construct buildings and provided a $6,000 grant. He also appears to have convinced the company's board of directors to contribute $4,000.[62] With the full support of the mining company, Menzel made the easy decision to place the coronagraph at Climax—merely 15 miles from Menzel's hometown of Leadville. Solar physics and sun–earth connection research had come to Colorado. Irrespective of Schott's motivation, and with WWII in its opening stages, Walter Roberts soon left for the great American West.

Roberts in the Rocky Mountains during the World War II

Driving west through Dust Bowl America in summer's heat stimulated Roberts's thinking on the sun–weather connection. His car's water pump succumbed to the heat, stranding Roberts and his new wife Janet near Columbus, Nebraska. Roberts sent both Menzel and Shapley telegrams requesting funds to fix the car and complete the trip. Shapley posted his telegram from the "Nebraska Refugees," as he called the couple, on the HCO bulletin board and labeled it Climax Bull. #1: "Colorado or Bust!—Busted."[63]

The newlyweds made it to Colorado, where Menzel met them in Denver for the hundred mile trip up to the observatory. Feeling impish, he took the couple to the ghost town of Kokomo and drove to the most dilapidated house he could find. He then introduced them to their "honeymoon cottage."[64] Prank done, they proceeded to their real future home at Climax. Roberts and his wife did not know the planned year-long stay in the mountains—long enough for Roberts to get enough research done to complete his PhD dissertation—would last seven years.

Roberts rebuilt the coronagraph, disassembled for the trip west, at Climax. The observatory hosted an open house, mostly for CMC employees and residents of nearby Leadville, on Sunday, September 8, 1940. Unfortunately, no one observed the corona that day. Optical problems with the coronagraph, combined with observing conditions that were poorer than

Menzel anticipated, combined to produce about a year's delay before the system generated useful images. Not until the autumn of 1941, and with the helpful Lyot and solar astronomer John Jack Evans, could Roberts claim a fully functional and operational coronagraph.

Both Roberts and Menzel had a strong sense of urgency in getting the coronagraph functioning properly and generating solar data. Not only was Roberts in a hurry to get the data he needed to complete his degree requirements, he also believed he raced against time as a young scientist. According to Roberts's son David, the twenty-six-year-old astronomer believed the notion that a scientist had to make his major discoveries by the time he was thirty.[65] Scientists of this era almost codified this view, called by physicists and historians the "curse of the Knabenphysik" (meaning postdoctoral or, literally, "boys' physics"). This idea arose because young men had made many of the discoveries of the new quantum physics. Albert Einstein had his "miracle year" when he was twenty-five, Niels Bohr developed his model of the atom at twenty-seven, and Werner Heisenberg was a mere twenty-three when he created matrix mechanics.[66] Roberts thought he felt the curse hovering over him as he closed in on his thirtieth birthday in the early 1940s and still had nothing significant to show for his scientific career.

Menzel, on the other hand, needed more funds to maintain the observatory and to pay Roberts's salary. Fundraising, as Menzel knew and as Roberts would find out, consumed much time, distracted from research activities, and often proved frustrating. A proposal to the Rockefeller Foundation for $10,000 did not fare well, for example—foundation officials, for unknown reasons, rejected it less than two weeks after it was sent. Given foundation director Wickliffe Rose's proclivity to "make the peaks higher," it seemed unlikely that the nascent, tiny, and remote solar observatory would garner foundation support. Still, by early 1941, Menzel had contributions from thirty-nine sources, with an average donation of about $200. Schott donated the most from this diverse list of contributors, with funding coming from the Department of Agriculture and Harvard Science Associates as well.[67]

The rapid onset of American involvement in WWII changed the fortunes of the tiny Fremont Pass solar observatory and the course of science in Colorado. As Menzel predicted to various sponsors over the years, the arcane coronagraph data had important practical uses. Wartime experiences as discussed below demonstrated this to a large degree and also temporarily ended the Climax station's funding struggles.[68] At Climax, Roberts demonstrated that solar research had national security implications and made discoveries that advanced the allied war effort.

Serving as the site's only observer, Roberts made a significant discovery at the end of 1941 that linked sun–earth connection science to military operations. After he accumulated enough data to do some detailed statistical analysis, he found that geomagnetic storm activity tended to occur a few days after he observed increased brightening over the east limb of the sun. This brightening occurred in a certain spectral line dominant in the sun's corona and visible spectrum—the coronal green line.[69] Electron transitions from within an iron atom ionized thirteen times produce this coronal emission line found at a wavelength of 530.3 nanometers.

These geomagnetic storms associated with this brightening affected the ability of radio waves to propagate in the earth's atmosphere—another important component of the sun–earth connection. The science had matured greatly because of the advances of solar observing and reporting in the mid-1930s and early 1940s.[70] As a result, Roberts stood ready to make his first notable scientific discovery. "In late 1941 I first noted," he wrote later for an American Institute of Physics career survey, "from my own data, a connection of strong green-coronal line emission to the short-wave radio reception quality."[71] As radio communications were rapidly becoming a vital part of military operations in the pre- and early WWII era, this was a most timely discovery. As a result, Robert's discovery of the coronal green line–radio reception relationship dramatically altered the fortunes of the fledgling observatory.

The Climax observatory, like other astronomical work in the nation during the war, soon underwent what one student of the period called a "general pattern of disruption and invigoration." Roberts's discovery provided the basis for this at Climax.[72]

Roberts started sending his coronal observations to J. Howard Dellinger, an ionospheric radio propagation expert at the Department of Commerce's National Bureau of Standards (NBS) in Washington, D.C. Dellinger was among the first to associate radio communication with solar activity.[73] It appears he noticed the same correlation between the coronal line and radio propagation at the same time Roberts did. According to Roberts's recollection, their letters informing each other of the discovery passed in the mail.[74] Coming at the cusp of America's entry into the war, Roberts's scientific work establishing this particular sun–earth link had practical import for communications. His data could help the nation and its allies predict the vital wartime ability to communicate over long distances. To add to the sense of urgency in gathering solar data, allied scientists feared that the Germans might have similar interests with respect to radio prediction capabilities.

Germany led the way in transforming solar observing into a military requirement. Radio expert Hans Plendl and solar astronomer Karl-Otto Kiepenheuer organized a Luftwaffe effort to observe the sun sponsored by a scientifically oriented general, Wolfgang Martini.[75] Joseph C. Boyce, at the newly created National Defense Research Committee and a friend of Menzel's, learned of the German ionospheric and solar interests from Norwegian astronomer Sven Rosseland in early 1942.[76] The concern over German ionospheric studies, coupled with the discovery of Roberts and Dellinger, immediately altered solar and ionosphere studies into subjects of high national security interest. It is important to note the uniqueness of Climax observatory's contribution to the war effort, for it helps to shape the future relationship between Roberts and the NBS. The war work also set the stage for some of the subsequent events in Boulder. "The contribution of our coronagraph station," Roberts wrote soon after the war ended, "has been to supply the only data available to the Allied Nations regarding the highly important corona of the sun."[77]

Menzel seized the opportunity and asked the NBS to have the Climax station declared a national defense project. Roberts, Menzel argued, ran the only working coronagraph in the western hemisphere available for allied radio propagation forecasting. Two other functioning coronagraphs existed, but they sat in areas beyond immediate allied control and use—Lyot's device in German-occupied France and another in neutral Switzerland.

Dellinger, whose operation at NBS quickly became a clearinghouse for geophysical data, approved the request to fund Harvard's Climax station in Colorado. By July 1942, Roberts operated under the full auspices of the wartime government.[78] "So important was this information," Roberts later recorded speaking of his then militarily sensitive solar observations, that "we . . . immediately coded these measurements for telephoning to Western Union in Leadville where they were transmitted with government message priority to Washington, D.C."[79]

After the government began supporting the Climax station, mostly via the NBS, Roberts did not have to worry about funding for the duration of the war. Also, as the superintendent and sole operator of the observatory until later in the war, he avoided being drafted in the armed services.[80] There existed some downsides to this turn of events. First, Roberts had to assume the burden of operating the observatory entirely on his own. Despite the perceived importance of his data in many defense circles, it was wartime, and the government did not post additional observers to Climax until 1944. Roberts had to make do, given the realties of wartime America. Few activities, even those declared of vital national interest, rivaled the importance of the Manhattan Project or the MIT Radiation Lab.[81] As a result of this inability to add staff, Roberts did not leave the isolated site for a stretch of eighteen consecutive months during the early part of the war.[82]

He dutifully took coronal observations every day the weather permitted, hoping to advance the war effort and his scientific career. Roberts later wrote of this time that he "was more or less frozen" in his job at Climax.[83] Even with having to suffer through difficult high-altitude winters, life in the mountains did have an upside for the growing Roberts family. Despite the rigors of life at Climax, and having three children in this austere environment in rapid succession, Janet Roberts reflected upon Climax as "a really good place, an ivory tower, literally, to spend the war years."[84]

One difficulty of the war years related to Roberts's doctoral dissertation. As Roberts wrote after the war, his work was "sufficiently important" enough to keep the data classified "until after V-J Day."[85] Roberts's dissertation was in effect "born classified."[86] Roberts later summed up the situation briefly by stating that he delivered the completed draft dissertation to Harvard, and "then the whole project was classified, and my thesis was classified."[87]

Irrespective of classification, Harvard awarded him the doctorate in 1943. Many understood at the time that much scientific research might surface only after the war ended. Even at that time, the American Astronomical Society's newsletter speculated that this phenomenon of wartime science did exist and that many important discoveries would likely surface after the war ended.[88]

Important professional relationships that determined the future of science in Boulder formed in the crucible of the war years. Because of the importance of radio propagation studies to the war effort, Dellinger created within the NBS the Interservice Radio Propagation Laboratory (IRPL) in 1942.[89] Roberts initially sent his data, via a "secret code," to John A. Fleming at the Department of Terrestrial Magnetism (DTM) at the Carnegie Institution, but once IRPL started routine operations at the NBS, the data went directly there.[90] This data flow and associated scientific relationship between Climax and IRPL continued for the duration of the war. This relationship had both scientific and significant personal dimensions. Menzel entered active duty as a lieutenant commander in the U.S. Navy and, as part of his wartime duties, worked with the IRPL on ionospheric forecasting and other radio propagation issues. In this capacity, perhaps in the tongue-in-cheek words of one biographer, Menzel "serendipitously aligned" the programs he had set in place at Climax and the NBS.[91] In reality, serendipity probably had little to do with these connections. Menzel may have already started thinking about postwar solar astronomy and the opportunities that might arise given the government's increased interest in sun–earth connection science fostered by the conflict.

In addition to a shared close connection with Menzel, the future presence of recent Harvard graduate and ionosphere scientist Alan Shapley, son of the HCO director and Roberts's mentor, further cemented the bond among the Climax operation, IRPL, and the Carnegie Institution's Department of Terrestrial Magnetism (DTM). The younger Shapley, at DTM during the war, joined IRPL's successor organization in 1947. Only a few years junior to Roberts, Roberts got to know Shapley at Harvard. The close relationships between Roberts at Climax and the IRPL staff, with Menzel as a middleman, shaped subsequent events in Boulder.

Roberts made other important contributions to solar studies in the process of his wartime solar observations and research that increased his reputation in solar astronomy and sun–earth studies. In addition to the possible relationships between effects on earth and coronal green line emissions, he also discovered that the corona itself rotates with the sun. This represented a significant contribution to solar studies since it conclusively showed that the corona is not a disjointed physical entity from the sun—it is directly coupled to lower solar layers. This discovery remains fundamental to the nature of solar structure.

The phenomenon Roberts came to call "spicules" represented his best-known discovery. Spicules are hot jets of solar hydrogen gas emanating from the sun's photosphere, in effect the layer of the sun visible to the naked eye. Roberts first observed these narrow jets near the polar regions of the sun, writing that in the "fall of 1943 I noticed that small chromospheric spike prominences were clearly discernable in photographs."[92] The very brief lifetimes and number of these features struck Roberts, and much of the paper relates to his statistical analysis of the phenomenon. Roberts recalled that, since they had the appearance of spike-like features, Harlow Shapley suggested the name spicules—literally meaning "little spikes." Shapley told his junior that, in referring to the phenomenon, "to be famous you have to have a name," and suggested the term Roberts eventually used in his discovery paper.[93] Roberts also sometimes charitably claimed that his observations just rediscovered the phenomenon. In his discovery paper he refers to nineteenth century Jesuit priest–astronomer Angelo Secchi's observations of "vertical flames" of the sun's polar chromosphere, a thin layer of the sun's atmosphere just above the solar surface.[94]

As the war came to a close in September 1945, IRPL contracted with HCO for Climax to continue providing solar data. This funding gave some measure of security for continued operations, but not nearly enough to expand the site's modest facilities and research agenda. Funding and sponsorship again played a vital role in determining the future of solar astronomy and sun–earth science in Colorado. Roberts and Menzel had ambitions to expand their scientific research. They decided that remote Climax was not a good location for these new efforts. As with many Americans facing the uncertainty of the postwar world, there arose an obvious question for Menzel, Harlow Shapley, and Roberts as to what the immediate future held for Harvard's Climax station and solar science. The answer pointed them to the city of Boulder, not far from Climax, and the University of Colorado.

CHAPTER THREE

From Leadville to Boulder

Some scholars describe the post-WWII years as a period of institution building in the evolving discipline of solar physics.[1] Boulder benefitted as it became a home for solar research and the closely associated study of the sun–earth connection. This chapter documents and analyzes how Walter Roberts unintentionally launched Boulder's growth as a science center. Roberts helped establish a particularly innovative institution, the High Altitude Observatory, in the immediate post-WWII years. HAO was the first corporation for science in the United States, as it preceded Associated Universities, Inc., manager of the Brookhaven National Laboratory on Long Island, by a few months.

This chapter also demonstrates how the incipient sun–earth science effort in Boulder faced a threat to its existence: Menzel and the Air Force's desire to build yet another new solar research site in the American Southwest inadvertently created a competitor for solar activities in Colorado. Luckily, Roberts and several fortuitous events precluded this potential turn that might have quickly ended Boulder's development as a city of science. These and other events also revealed a split between Roberts and Menzel on how best to fund science—a split that in many ways reflected a broader divide in American astronomy on how to fund modern astrophysical and associated research.

Roberts and Menzel, because of their wartime experiences and the alliances they formed at that time, understood that the time was ripe for the creation of new facilities and organizational arrangements in U.S. astronomy.[2] Menzel and other colleagues at Harvard, such as Fred Whipple in his upper atmosphere studies, saw opportunities to undertake innovative research and associated infrastructure building by securing government funding.[3] Having seen the benefits of large-scale science support from the government during the war, they had no desire to return to the prewar status quo.[4] The advisability of using these funding new sources, however, produced tensions at Harvard and throughout the broader astronomy community.[5]

Choosing to rely less on the research and funding decisions of an observatory director by exploiting new government funding, Menzel and others quickly began to exercise their own authority after 1945.[6] This new strategy is essential to understanding the development of sun–earth science and postwar developments in Boulder, and it signaled the beginning of the end of the imperial observatory director tradition in U.S. astronomy. It also enabled astronomers to develop their discipline beyond its traditional emphasis on stellar and galactic research. Solar research therefore underwent a period of significant expansion in the postwar years, and Boulder benefited from this expansion.[7] Observatory directors such as Shapley and others began to lose what astronomer Donald Shane called their "freedom of action" as new funding sources allowed younger scientists to seek their own scientific fortunes.[8] As a result, Shapley's "divide and rule" tactics—he often put subordinates at odds with each other, including Roberts and Menzel—became increasingly ineffective.[9] Even Roberts, who had great personal affection for the HCO director, thought Shapley "devious and arrogant" at times.[10]

Menzel and other scientists understood, to varying degrees, the significant changes occurring in the landscape of U.S. science funding. They also understood how these changes might increase their research autonomy. Many scientists therefore eagerly sought out, in creating new scientific institutions, ways to take advantage of these new sponsorship patterns arising from the U.S. government's increasing interest in funding scientific research. Roberts moved to Boulder in 1946 to help create one of these institutions enabled by these shifting patronage patterns. This was a new type of corporation of associated universities, one for solar science.

An Innovative Arrangement: A Science Corporation?

The idea to transform the HCO Fremont Pass Station at Climax into a joint Harvard–University of Colorado endeavor originated in Menzel's desire to have stronger ties to the state of Colorado in order to enhance his possibilities for fundraising by increasing local political support and patronage. This approach formed part of his strategy to gather support for his original coronagraph operation, but nothing ever came of potential relationships with the University of Denver, Menzel's alma mater, or Colorado College in Colorado Springs.

It appears that HCO director Harlow Shapley resurrected the idea of a joint project in October 1944 and proposed the University of Colorado at Boulder as a candidate. He thought the school "much stronger and probably more highly respectable" than other candidates such as the University of Denver. Shapley suggested that he could help Menzel "very carefully and cautiously feel out the situation" with University of Colorado president Robert L. Stearns.[11]

Menzel responded that he had "actually been thinking along the lines" Shapley had proposed. He agreed "that Colorado is a stronger institution all the way around" in reference to other schools. But always conscious of funding issues and not offending potential donors, he prefaced this remark with "please don't quote me." By staying in the state, there still might exist an opportunity to tap into the "Penrose millions" in Colorado Springs, even if they did not approach Colorado College to participate in the joint project.[12]

Stearns was on wartime leave from the university at the time and serving as a civilian advisor to the U.S. Army Air Forces in Washington, D.C. Shapley contacted Stearns, and the two met for lunch at New York's Century Club in late November 1944. Stearns liked the idea of a joint project because the partnership might associate his university with the Harvard College Observatory in Climax, already a noted astronomical research facility, and forge a research link with the prestigious Harvard University. Stearns accepted the proposal while expressing the desire to assist in fundraising for the cooperative effort. Stearns wrote to the university's acting president, Reuben G. Gustavson, asking him to raise the idea with the school's regents as he thought this relationship would greatly benefit the university.

Shapley reported to Harvard president James B. Conant about a meeting Shapley, Stearns, and Menzel had in March to discuss the collaboration further. The HCO director reported to the president that Menzel felt "quite confident" that funds in excess of $200,000 might come from Colorado donors to cover HAO start-up costs and that the project should therefore cost little or nothing in existing Harvard research funds because of the desire of local donors to improve science in Colorado. The donors also wished to associate these new Colorado activities with Harvard. Shapley added that the project "would be scientifically of high value to Harvard" and of "high publicity value in Colorado." He concluded that the project was therefore well worth pursuing on both ends.[13] Convinced by Shapley's arguments, Conant assented to continued discussions on the matter.

Momentum picked up on the Boulder–Cambridge project as Stearns returned to Boulder and discussed his desires to proceed with the joint arrangement with the University of Colorado regents. The regents apparently came to understand the potential benefits of this relationship to the University of Colorado and authorized Stearns to investigate the joint management concept.

In September 1945, J. Churchill Owen, an attorney with a Denver law firm employed by the university, wrote Stearns that Harvard and the University of Colorado could, and should, form a nonprofit corporation since Harvard had expressed a desire that neither university own the land and facilities outright. The entity, even though a corporation for science alone, would have a board of trustees to oversee its activities. Owen proposed the board consist of three trustees appointed by each university. The Denver lawyer had planted the seeds for an astronomy corporation composed of universities, the first such entity in the country.[14]

Owen, a tax and corporate lawyer, had unknowingly hit upon an idea that high-energy physicists from a number of prestigious universities in the northeast arrived at as well—to organize modern U.S. science collaboratively in the form of a corporate legal structure. Called the Initiatory University Group, the physicists patterned their new nuclear physics lab on the contract structure for Los Alamos, but with many universities running the lab via a consortium. Groups could associate legally in this manner to share the burdens, and gains, of establishing such research entities. The universities involved in the discussions incorporated under the Education Law of the State of New York to form Associated Universities, Inc. (AUI) in July 1946 to oversee the still existing Manhattan District–funded Brookhaven National Laboratory.[15] This trend of forming corporations for academic science accelerated in the 1950s as the then-new National Science Foundation enthusiastically began to fund such entities, and the turn to academic corporations played yet again a crucial part in Boulder's development in the late 1950s.

The incorporation technique afforded benefits to all participants, and by the 1950s it became commonly used as universities increasingly wished to pool resources for scientific endeavors. It allowed multiple preexisting organizations to pool resources legally and to simultaneously seek both public and private funding. An operation owned by the government could not use anything but taxpayer dollars, except given a specific exemption by law, such as what the Smithsonian received. As HAO was a private institution, it was free to use funding from public or private sources. For example, both the U.S. Navy's Office of Naval Research (ONR) and the Research Foundation, a private philanthropic organization, funded HAO instruments. In addition, the arrangement enabled scientists to retain more autonomy in their own operations, an important and decisive issue for many scientists. They could obtain significant funding from the government, yet they did not have the same constraints as an in-house government lab might. This ability to obtain funding from both the government and private sponsors played an important role in the continued existence and flourishing of HAO in the late 1940s and 1950s.

Encouraged by Owen's positive response, Stearns arranged for Shapley to give a presentation to the regents on September 21, 1945. A large dinner gathering followed on September 24 that included not only Shapley and Roberts, but also many local Harvard and Colorado alumni and others "who might be interested in . . . raising money for this project."[16] Following Shapley's talk, Stearns told the regents that "progress had been made" on this project and "that much of the laboratory work and the compilation of data should be done on the Boulder campus." He added, as might be expected based on his high ambitions for his school, "that the present laboratory facilities in physics [sic] are inadequate."[17] This led to a broader discussion of overall university building requirements. The university president steered the regents toward his vision of a new, expanded science effort, with the Harvard project proving a significant stimulus in this regard. Stearns was the first in a series of University of Colorado presidents who aspired to expand the school's research programs.

As a result of Stearns's acceptance of Shapley's suggestion, the idea of this new corporation for astronomy was "a major development in the history of the sciences" at the University of Colorado—there existed little research in astronomy there and, relative to major research schools of the time, little physics as well.[18] Possibly because Roberts had developed good relationships with both Stearns and his wartime fill-in Gustavson, Stearns sensed some opportunity to bring additional science activity to the campus.[19] Future events proved him prescient in this regard. Salaries for the project came from Harvard and operating expenses from gifts and government funding.[20] The move would therefore produce huge dividends for both the school and Boulder because the new relationship to Roberts and his sun–earth connection research set a foundation in Boulder for more space and atmospheric science research activities in the ensuing years.

Development of the observatory proceeded quickly. Initially, according to Roberts, they wanted to have "Climax" in the title. However, Climax Molybdenum Company had a legal claim to the word "Climax." In a rush to come up with something else, Roberts recounted, they thought that since the observatory was the highest in the world at that time, a reasonable interim name was the "High Altitude Observatory."[21]

The new joint Harvard–Colorado solar observatory began life as the High Altitude Observatory of Harvard University and the University of Colorado. Officials filed the incorporation papers under the laws of Colorado on April 12, 1946. The initial governing board of the nonprofit corporation consisted of six trustees, three representing Harvard and three representing the University of Colorado.

Showing the growing enthusiasm among the populace in bringing modern research to the area, none of the individuals on the original HAO board, unlike the AUI board, were scientists or academics. All board members were prominent citizens of the local Denver–Boulder region. William S. Jackson, Peter H. Holme, and Erskine R. Meyer worked as lawyers practicing in the state. William C. Sterne, director of the Public Service Company of Colorado; E. Ray Campbell, publisher of the *Denver Post*; and Earl L. Mosley, city manager of Colorado Springs, rounded out board membership.[22]

The board members expressed both amazement and amusement about a corporation created for astronomy and scientific research, which is not surprising. They most likely did not know of the developments associated with AUI's creation, and that HAO stood right at the leading edge of the new wave of post-WWII scientific institution building in the United States. It is also unlikely AUI member's knew of HAO's creation, a much smaller effort and distant from the northeast.

The unusual arrangement of a corporate legal status left the HAO board baffled as to how best to describe this new entity of scientific research. Perhaps reflecting a natural tendency for the use of Colorado wildlife metaphors, founding board member William Jackson, a Harvard graduate and member of the Colorado Supreme Court, recalled his initial thought that HAO was a "strange beast" and that one of his fellow board members liked to ask, "Have you ever seen an animal like this before?"[23]

To assist the non-scientifically trained trustees in their deliberations, Harlow Shapley suggested the HAO board create a Committee on Scientific Operations (CSO) for the HAO. Shapley, Menzel, and Roberts were the CSO's first members.[24] Although a CSO seemed a useful entity at the time, disputes concerning the proper role for the CSO became one of the many fissures that led to the breakup of the arrangement in the 1950s. Both groups served important purposes for the fledgling observatory, but the issue of who really controlled the corporation's activities—the CSO in Cambridge, or Roberts and the HAO board in Colorado—loomed larger as time passed, leading to the eventual dissolution of the corporation in the 1950s.

This innovative corporate arrangement therefore let Menzel and Roberts seek funding from almost anyone they wished. Yet, this robust funding ability helped create a rift between the two men. Although Roberts did not share Shapley's suspicion, if not disdain, for military funding of science, Roberts thought it very important to find private funding sources that could provide him with increased and guaranteed latitude in research he might wish to pursue. Menzel wanted full government funding, irrespective of whether the source was part of the Department of Defense or not. This bifurcation of views about the proper role of government in science sponsorship polarized the 1950s astronomy community and led to and exacerbated personal friction between Roberts and Menzel.[25]

A primary sponsor of the new High Altitude Observatory was the successor to the WWII-era Interservice Radio Propagation Lab, the National Bureau of Standards' (NBS) Central Radio Propagation Laboratory. The new civilian lab evolved directly from the IRPL, one of Roberts's primary customers (and sponsors) of the observatory's solar data during the war. Menzel, working with Edward U. Condon, noted physicist and director of the NBS from 1947 to 1952, and J. Howard Dellinger of the Ionosphere Lab, established the requirements for a center that could ensure that the ionospheric forecasting and geophysical data services provided to the nation during WWII continued.[26]

The creation of CRPL on May 1, 1946, only a few weeks after the observatory's inception, thus ensured a continuation of the observatory's primary customer and a crucial source of HAO's funding. For the observatory, the Central Radio Lab's creation ensured continued basic operations into the immediate postwar years—the U.S. government, via the lab, needed and was willing to pay for the data. For Boulder, the significance of the lab's creation rested in setting the stage for the D.C.-based Central Radio Lab's relocation to Boulder less than a decade later.

There were not only strong professional connections among the HAO, the University of Colorado, Harvard, and the National Bureau of Standards, but a series of continuing and evolving close personal links as well. This social chain included Harlow Shapley's son and Roberts's friend, Alan Shapley, a key CRPL scientist and administrator. The younger Shapley played a prominent role in helping making Boulder a site for sun–earth connection research. Indeed, personal and professional relationships played an instrumental role in bringing space and atmospheric science to Boulder and were fundamental to Boulder's scientific creation.

By creating HAO, Harvard and the University of Colorado hoped to generate new intellectual, research, and funding opportunities. The association between Harvard and the University of Colorado also had another important effect, whose consequences could not be foreseen. It brought Walter Roberts to Boulder.

Roberts Relocates to Boulder in 1946

Roberts, after years in relative isolation, needed new outlets for both his scientific curiosity and personal ambitions. The mid-1940s turned out to be a time of internal questioning for Roberts. His wife, Janet, told her son, David, years later that Roberts's thirtieth birthday in August 1945 "was an occasion of agony for him."[27] He thought the important discoveries he made about the sun with the coronagraph during the war years did not put him on the top rung of U.S. astronomers that his mentors, Menzel and Shapley, occupied. He was an ambitious young scientist, and he believed his productive years were passing by without achieving lasting scientific fame.

No doubt the sense of isolation from the academic mainstream produced by the years at Climax played a role in creating the "agony" that science had passed him by. With the war over, Roberts decided to seek a more collegial and intellectually stimulating environment than he could find on a mountaintop. Not wanting to totally separate himself from solar observations at Climax, nearby Boulder seemed a natural place for Roberts to grow intellectually and perhaps make a major discovery based on his solar work. Why did he move to Boulder and not some other location in the state, such as cosmopolitan Denver?

The University of Colorado at Boulder exerted both an intellectual and cultural pull on the academic Roberts. The university was certainly not Harvard, nor was Boulder a Cambridge. The school did not even have an astronomy program. However, the university's mere presence attracted intellectual and scientific activities to Boulder after the war. The university was good enough for Menzel and Roberts's specific, short-term purposes to bring solar astronomy (and sun–earth connection science) west. They wished to create a theoretical astronomy group at the school as they planned to expand the intellectual horizons of the solar observatory, in effect benefiting both their operation and the university simultaneously.[28] Stearns had similar ambitions and saw Roberts's association with the school as an important step in enhancing the university's reputation.

Roberts had already started making connections at the university by giving a few lectures in 1945 on science and society, a subject of lifelong interest to him.[29] In May 1946, soon after Harvard and Colorado began their corporation for astronomy and upon Stearns's request, the physics department put Roberts on their faculty. Roberts became a part-time research associate in astrophysics with an annual salary of $1000.[30] About the same time, as part of the Harvard–Colorado agreement, HAO's administrative offices relocated to Boulder.

As HAO's organizational structure evolved, Roberts took a long-prearranged sabbatical at Harvard for the academic year 1947/48 to refresh both himself and Janet intellectually and renew family associations. After seven event-filled, productive years that lasted six years beyond initial expectations, his mountain sojourn ended. When Roberts's eastern sabbatical ended in 1948, he returned to Boulder, not Climax, to set up his home and operations for good. As Roberts later said, it was at this time that "we started the real expansion of the High Altitude Observatory."[31] It was not just HAO's expansion that started at this time, but the beginning of sun–earth science in the city as well. What sort of city did Roberts and Janet find following their sabbatical year at Harvard? Boulder's history was linked to the natural environment, and the sun–earth connection science coming to the city through Roberts's efforts fit well into this theme.[32]

Boulder, circa 1948

Founded in 1859, Boulder's fortunes depended primarily upon mining. Toward the later part of the nineteenth century, agriculture, tourism, and higher education in the form of the University of Colorado determined much of the city's development. Located about forty miles northwest of Colorado's new capital, Denver, Boulder appeared in 1861 as the planned site for the future University of Colorado through the efforts of New England transplant Robert Culver, owner of a Boulder quartz mill, and Charles F. Holly, a local politician.[33] Boulder citizens at this time banded together to bring modern learning to their small town. To stimulate the state legislature, they developed the slogan, "Give Boulder the State University, and the rest of Colorado may take all other institutions."[34]

After much wrangling, construction began in 1875. Although mining eventually ceased, the university remained in Boulder and quickly became an important presence. Agriculture soon took the place of mining in the local economy, with strong support from tourism. The latter included a bit of what some refer to as "medical tourism," primarily related to the search for tuberculosis cures.[35] Boulder became a place to visit not only for the beautiful scenery, but also for the supposed health benefits of high altitude and nearness to the mountains. Despite this temporary influx of visitors, Boulder citizens did not encourage further settling, following the advice of noted American designer of outdoor space, Frederick Law Olmstead, Jr. In a 1908 report, Olmstead thought Boulder should develop primarily as a residential community in order to retain its scenic environment.[36]

As Boulder became a scenic preserve, the city's non-student population remained nearly constant in the early part of the twentieth century. Between 1910 and 1950, the residential population increased by about only 33 percent.[37] In the same period, the university student presence increased by a factor of almost six, from nearly 1,300 in 1910 to over 8,000 in 1950, with the greatest growth in the 1940s. This growth spurt reflected the wartime expansion of the school, coupled with increased enrollments in the postwar period.[38]

Much of this expansion occurred during University of Colorado president Robert L. Stearns's tenure. As evidenced by his enthusiastic reception of the joint Harvard–University of Colorado solar observatory, Stearns, although not a scientist, understood the importance of increasing the university's reputation by becoming more involved, as an institution, in leading-edge scientific research.

Stearns, a 1914 graduate of the university, attended Columbia University for a law degree and then developed a reputation as a legal educator. Stearns's predecessor, George Norlin, had a great vision for the university as a regional center of knowledge, and Stearns, a politically savvy administrator, began to fulfill that vision using the opportunities presented by increased federal spending in the latter years of the Depression, WWII, and the immediate postwar period.

His relatively long tenure as president, 1939–52, occurred during a period of much turmoil and change on both the national and international scene, and the events of his tenure often reflected the events of the period. He described the context of this tenure as "epochal."[39]

Stearns took time out to work as a high-level civilian operational analyst in the Pacific for the Army Air Force in WWII and a consultant for the U.S. Air Force in Korea. As was the case with many academics who participated in the national defense effort in the war, Stearns's duties served to increase his awareness of government opportunities and establish high-level government contacts that could assist him in implementing his ambitious plans for the university.[40] He established HAO with Harvard as one attempt to develop his university's reputation for academic research.

Stearns enlarged his university's faculty in all areas and both revamped and expanded the curriculum on a broad front by adding new basic courses, in addition to a vigorous program of building and facility construction facilitated in part by the burgeoning former GI student population and national training requirements during the war.[41] His association with Roberts and Harvard's solar observatory in the nearby mountains fit his larger vision. This association had more of an effect on the university and Boulder than Stearns could have then imagined, for it provided additional impetus for Roberts to remain in Boulder.

Sacramento Peak Solar Observatory—Complement or Rival?

Roberts played a crucial role in keeping Boulder on a path to become a center for science. There was a move in the summer of 1947 to shift the Harvard solar observing operation from Climax to a new site in New Mexico. The loss of HAO would likely have altered the city's subsequent fate in fundamental ways. Roberts intervened to ensure incipient sun–earth research continued. Even though Roberts was a coauthor, along with Menzel, in creating a proposal for the new solar observing station somewhere in the U.S. Southwest, he certainly did not conceive of it as a potential threat to Boulder or Climax operations.

Menzel, no doubt stimulated by his wartime activities and the success of creating HAO, expanded his astronomical activities in the west when the Air Force offered support for the next addition to what Menzel eventually referred to as "his western solar stations."[42]

After the Navy declined his advances about building such a facility, probably because of the sheer size and cost of the operation, Menzel found a receptive, and enthusiastic, ear with Marcus O'Day of the U.S. Army Air Forces' (AAF) new Cambridge Field Station. The AAF (redesignated on September 18, 1947, as the U.S. Air Force) had a rapidly evolving interest in the upper atmosphere and outer space. Important Air Force leaders, such as General Henry "Hap" Arnold and his science advisor, Theodore von Kármán, saw these regions as important to the service's future operations, given the advent of missiles and high-flying aircraft. Understanding the physical nature of these regions was therefore a high priority for the service.[43]

O'Day was at the forefront of these interests, and his Cambridge group found itself as the lead in the Air Force's effort to use captured V-2s for upper atmosphere research.[44] O'Day suggested to Menzel that the Air Force might fund solar research as an augment to the V-2 studies.[45] The question soon became as to how O'Day, Menzel, and Roberts could meld their disparate but overlapping research desires into a coherent package of activities that the new Air Force would support.

The U.S. V-2 research rocket activities centered around the White Sands Missile Range, about forty miles southeast of Alamogordo, New Mexico, and fifty miles east of the Sacramento Mountains. These operations served as a magnet for scientific research that could both complement and augment the V-2 effort. In addition to Menzel's operation, the V-2 research also attracted Menzel's HCO colleague, Fred Whipple. Whipple relocated the Naval Ordinance and Naval Research Lab–funded Harvard meteor research program from Flagstaff, Arizona, to Las Cruces, New Mexico, precisely to be close to the White Sands area, in addition to seeking better observing conditions.[46]

A key requirement for locating the new observatory was the ability to have total surveillance of the V-2 flight path so that personnel could closely couple solar observations and rocket launches.[47] The general collocation of the launch and observation sites enabled scientists with instruments on the rocket and on the ground to better understand their results. Conversely, solar observations might inform launch personnel on the best time to launch rockets for this data collection.

O'Day, in particular, wanted a solar facility in the vicinity of the rocket launches, in part so he could do this type of close coordination with an advanced solar observing facility.[48] "We had to have line of sight to the ground stations," and the site chosen also "had to have full surveillance of the range," Roberts later recounted.[49] These criteria restricted the possible locations for a new observatory to a site somewhere in the Sacramento Mountains to the east of White Sands.

Roberts initially displayed little enthusiasm for Menzel's idea of a "New Mexico station." He wrote to Shapley in July 1947 that he was going with Menzel to visit potential sites and was "far from sold on the idea."[50] His initial reluctance most likely arose from the simple fact that he had much work to do in setting up the Boulder operation and carrying forward the plans for Climax. He may have thought that the new site might drain both time and resources from Boulder.[51]

Once Roberts got to the Sacramento Mountains, the location's excellent observing conditions "vastly influenced" Roberts's opinion on the advisability of the project. This reaction reflected a common sentiment among those visiting the Sacramento Mountains. Fellow astronomer Leo Goldberg journeyed to the proposed site in the fall of 1947 and reported back to Shapley that "the site was wonderful" and that he had never "seen a sky so black near the sun."[52] Roberts supported Menzel's plan, stating to Shapley that the "establishment of a south station now looks to me like a *very* desirable thing."[53]

For Roberts and Menzel, then, the proposed observatory could offer a host of new opportunities as a complement to their existing observatory at Climax, despite a possible negative effect on the Colorado operation. As Roberts wrote Shapley, a new observatory would "multiply our Climax station value by far more than a simple factor of two." Roberts overcame lingering concerns, and assuming funding could support both sites, he came to fully support the new effort.[54]

In contrast, some astronomers at the Harvard College Observatory (HCO) began questioning the wisdom of having a growing number of HCO research outposts so far away from Harvard. These concerns apparently centered on the logistics of maintaining new solar stations in the American West, in addition to Whipple's meteor site. Other concerns related to further drains on staff time in Cambridge in administering these new sites.

The suggestion evolved to a point where Shapley even wrote to Roberts in late July 1947 with a list of "for and against arguments" relating to the New Mexico station. Much of the discussion about the new Air Force–sponsored observatory reflected not only Shapley and the HCO council's concerns, but also reflected prevailing anxieties facing U.S. scientists and science policymakers of the time.

Among Shapley's concerns were "devotion to instruments and equipment rather than publishable analysis" and "getting too directly under Army control (security regulations, red tape, bad company, etc.)." Some scholars, perhaps most notably Paul Forman, have written about the turn toward the instrumental and "techniques" (and away from the theoretical) as a concern about a perceived result of military funding in particular.[55] Shapley's memo reflects this, although his thoughts also reveal more of his thinking about science–military relationships.[56]

The thoughts of Shapley and Menzel represent a spectrum of prevailing views on the advisability of government (and military) funding, with Roberts somewhere in the middle. There appears to be a generational issue involved in this question as well as possible ideological and control issues, including Shapley's well-known pacifism and his iron grip on HCO funds. The much younger Roberts and Menzel understood, as did many of their peers, that questions of modern science required new, more complex (and, as a result, much more expensive) instruments unlikely to garner sponsorship from traditional sources.[57] They also understood that dependence on the willingness of imperial directors such as Shapley to sponsor research not in their own fields might lead to great frustrations, as it had when Shapley told Menzel in the 1930s to go find his own money for solar work. The split between Roberts and Menzel did not arise over a debate on if the government should fund science—they both agreed it should. The issue, rather, was what mix of civil–military and government–private sponsorship best suited the modern scientific enterprise.

"The possibility of witch hunting involvements (classified Army work)" Shapley thought merited only the lowest level of concern.[58] Some scholars refer to the widespread deleterious effects of classification on scientific discourse and claim that "strictly speaking there was in this period no such thing as unclassified research under military sponsorship" because the mere possibility existed that the military could have classified the work.[59] Yet, irrespective of this idiosyncratic definition of what "classified" actually means, little postwar solar research became classified. It was precisely the free flow of sun–earth connection research and coordination among researchers, both in the city and beyond, that played such an important role in Boulder's development as a science center in the 1950s.

The out-spoken Shapley's belief that "witch hunting" in mid-1947 warranted little concern is surprising given the generalized anxiety of the time produced by loyalty oaths and security hearings. His liberal opinions often put him at odds with the political establishment in this era, including significant Harvard alumni donors.[60] Perhaps most controversial of all were his views on scientific internationalism—he thought that the sort of cooperation that existed in science could exist among nations as well.[61] As a result, he often criticized what he viewed as unnecessarily bellicose U.S. Cold War policies relating to the Soviet Union.[62] Shapley particular resisted the charge that he was pro-Soviet in doing this, sympathetic to totalitarianism.[63] Proud, even arrogant in his defiance, Shapley referred to the House Un-American Activities Committee as the "Un-American Committee."[64] Almost any association with Shapley could put colleagues under government suspicion.[65] This outspoken approach therefore produced conflict at the HCO. Ironically, the New Mexico project inadvertently occasioned a security clearance controversy in 1950 for Roberts.[66]

It is important to note of the fourteen items "against" the New Mexico solar project, only two had the most negative effects of the project. These included Menzel and others' time at "the largest graduate school of Astronomy ever," and "the neglect-of-family angle" due to the distance from Harvard and Boulder.[67] Logistics and practicality trumped all other issues, even in Shapley's mind.

Of the eight "for" arguments on Shapley's list, only "the strengthening of our solar work" and "the possibility of adding L.A. to the staff," referring in particular to astrophysicist Lawrence Aller, received his highest rating. Both clearly worked, even in Shapley's hesitant mind, in favor of establishing the new observatory with Air Force money.

Reasons of lesser importance Shapley gave to proceed with the effort included the enhancement of HCO prestige, "the association with 'easy' money," and "greater national service" by working with the Army, the Weather Bureau, and the NBS.[68] These thoughts illustrate that Shapley had at least conflicting views about military support and associating science with national security. But Shapley's views did not interfere with him seeing "easy money" for science as a "positive" for HCO, irrespective of the source. These "for" arguments he gave Roberts to ponder shed light on Shapley's thinking at the time and demonstrate he may not have had the absolute views often attributed to him by scholars.

Roberts responded by writing that he "almost completely" agreed with Shapley on the list, adding that he shared both Shapley's "worries" and Menzel's "enthusiasms."[69] In a quick turn of events, the HCO council's eventual decision to proceed with the New Mexico project raised for a time serious questions about the future of the still developing HAO—Shapley related to Roberts on August 21 that to his "surprise" the council wanted to closely examine a "transfer of the Climax station to the Alamogordo site."[70]

Among the reasons Shapley gave for considering the move away from the nascent Colorado operation, funding loomed large. Failure to obtain substantial support from "Colorado interests" topped the list, for some of the hoped-for support from local sources did not materialize after HAO's creation. Other reasons, echoing Shapley's previous letter on the creation of the New Mexico site, included "the avoidance of the questionable practice" of creating HCO-run "out-stations" in the far West and the failure to obtain funds from the Army, Navy, or other sources for major building construction at the planned new site at Climax.[71] The HCO thought that Climax might not pay its own way, for private funding did not materialize easily for the facilities to house the military-funded instrumentation.

Shapley added "a simple explanatory letter" to the HAO trustees on "the inadvisability of continuing . . . at our present site," and the problems of getting funds for a new site "should suffice to convince the Trustees of HAO that for the good of science the organization should be gradually dissolved."[72] Events, and the possibility of large new amounts of government funding for the newer, larger site in New Mexico, put what Shapley previously had called "our lovely Colorado project" on the chopping block.[73]

"How much embarrassment such a move would cause at Boulder . . . I cannot predict," added the HCO director. "How much this would upset your own personal position and plans, as a Colorado citizen, also must be considered." Roberts and a recently added scientist to the Climax staff, Harvard-trained solar instrument designer Jack Evans, needed, therefore, to provide their opinions.[74]

Roberts, no doubt shocked by the possibility of losing his new Colorado operation to New Mexico, responded. He understood that he was on the verge of losing all the work he had put into building Climax and the newer Boulder operation. His firm response to Shapley established the foundation for many of the future developments in Boulder with respect to space and atmospheric science, for it kept this nascent scientific community viable in the city. This episode also revealed Roberts's desire to retain as much autonomy as possible for the scientific enterprise by not getting closer than necessary to government sponsors, even if it required continued hunts for private funding of his activities.

"Evans and I do not think this will be a feasible thing," he wrote on August 26, referring to Shapley's proposal to slowly dissolve Climax work and shift operations to New Mexico. He gave the isolation of the proposed site (as compared to Climax), the "extreme unpleasantness" of temperatures in Alamogordo, uncertainty still about site observing conditions, "the absence . . . of a campus atmosphere," and "the undesirability of a very close relationship with the Army or military in these research programs" as reasons for not supporting the proposal to shift all solar work to New Mexico.[75]

This latter concern appears because the original concept for the Sacramento Peak site centered on its direct support of V-2 efforts in the nearby White Sands and, unlike Climax, might exist only to support such work.[76] Roberts, again, showed himself on the spectrum of views between Shapley and Menzel with respect to the desirability of government funding of astronomy. He eagerly accepted funding from the military for his Climax/Boulder work, but he did not want to run a government-owned facility.

Roberts added in a separate paragraph perhaps the ultimate and most frank reasons for not moving. He wrote that "it would be of utmost difficulty to keep our respective spouses . . . in a state of happiness in . . . a place like

Alamogordo, Las Cruces, or El Paso." Alamogordo, in particular, "is still a dirty and backward town the way only cities near the Mexican border can be."[77] Roberts added that "we are dominated, I suppose, as much by the reluctance to settle in a location like Alamogordo, with heat, sordid surroundings, and rattlesnakes." One way or another, then, "if Alamogordo (Sacramento Peak) becomes our main station . . . we must also expect that there will be phycological [sic] disadvantages" requiring frequent rotation of personnel back to Cambridge.[78] Roberts and Evans (and their spouses) did not want to move from Boulder's intellectual and cultural opportunities provided by the presence of the University of Colorado.

In handling this threat to HAO's continued existence in Colorado, Roberts suggested a compromise of sorts. In order not to enter a head-to-head competition with the new site in New Mexico, he suggested each site have a suite of instruments and staff that complemented each other. Personnel at Climax and Sacramento Peak could work then in effect as a team and not as rivals.

To attain these goals, he proposed that the "best arrangement" was to "not build up anything beyond minimum necessity at Climax." The "minimum" included one of the "super" coronagraphs then in design or an improved single-tube coronagraph for coronal work only. In addition, he proposed for the new Climax observatory a "small scientific community" including a cosmic ray laboratory and about two or three times the existing space at Climax. The Climax site, in Roberts's view, could exist primarily for special research projects while the New Mexico operation would focus on routine observations and government research. Climax, in this manner, might serve both as a backup for and adjunct to the New Mexican activity.[79] Menzel appears to have agreed with Roberts's view, as he did not actively support the idea to do away with HAO by combining it with the new location.

Roberts also suggested that the telescope recently donated to Harvard by the Bausch and Lomb optical company go to Boulder, in addition to a small laboratory with office and other spaces, including a machine shop. These facilities would clearly strengthen the scientific tie between the HAO and the university and serve to further deepen HAO's roots in Boulder.

These discussions continued in the fall of 1947 after Roberts returned to Cambridge on his sabbatical. As late as December 1947, Shapley, in a memo to Roberts and Menzel, still left open the possibility of a move of Harvard's Colorado operations to New Mexico.[80]

The New Mexico project advanced significantly in December when Menzel and Roberts submitted a recommendation to the Air Force to build the new observatory on Sacramento Peak.[81] In May 1948, the Air Force Materiel Command issued a five-year contract to Harvard with Menzel as the principal investigator for the construction of the observatory colloquially known both then and now as "Sac Peak."[82]

The contract for Sac Peak, however, did not spell the beginning of the end for the Climax and Boulder operations. The Air Force and the Navy about this time also agreed to fund the cost of developing both new coronagraphs as part of their burgeoning interest in the space sciences, one at Sacramento Peak and the other at Climax. This new agreement obviated the funding concerns and there were no further discussions about closing the Climax site or dismantling HAO. It appears that by spring 1948, Roberts's compromise plan offered the previous August had saved HAO from being terminated. HAO continued operations in Boulder and Climax. Roberts, Evans, and others then served as important guiding hands in building Sacramento Peak while they also built the new, enlarged Climax site. As a result of this parallel development, close working ties and excellent professional relations existed between HAO and Sac Peak for many decades.[83] Roberts's compromise of complementary programs at the two facilities worked. It also ensured solar research (and thus sun–earth connection research) stayed in Boulder, a *sine qua non* for the city's future development as a center for science.

Funding Solar Science—Differing Views

This period of Roberts's activities in Colorado and the creation of Sacramento Peak also revealed the growing split between Roberts and Menzel on how best to fund solar research in the immediate post-WWII era. Their split mirrored a wider one in American astronomy.[84] One group, including many observatory directors, wished to continue more in the traditional vein of private- and state-funded support for astronomy.

Shapley did not want to take direct industrial or military support if he could avoid it. Speaking for himself and the like-minded, Shapley reported to Roberts at the time that "there is much worry among some scientists about getting under Army control." He added, perhaps thinking of Menzel and Fred Whipple of the HCO, who aggressively sought such funding, that "others do not worry about it."[85]

The group consisting of Menzel, Whipple, and younger astronomers such as Leo Goldberg, however, was eager to accept governmental funding for either HCO or personal research, even if some of this work fell into classified areas.[86] Certainly the latter group rode a tide of funding—the Office of

Naval Research, for example, provided a significant majority of support to basic science by 1946.[87]

Alvin G. McNish, who later that year joined the Central Radio Propagation Lab site selection team, perhaps best articulated the view of these "others" to whom Shapley referred. Writing to a friend at MIT in January 1949, he opined that "we should stop kidding ourselves with the belief that the military has any real interest in basic research any more than they have a real interest in the Metropolitan Opera." He added that he would not like to have the military establish "the final justifications on what is good basic research," though they did "appreciate what research and development, including basic research" contributed to national defense. As indicated by the creation of the Research and Development Board of the Office of the Secretary of Defense in 1947, "they have shown a willingness to ask for guidance in this field from those who are most able to guide."[88] Many scientists accommodated themselves to military and broader government funding as long as they thought they had sufficient voice in the process. In this postwar era, they had gained influence in determining research agendas despite the military patronage.[89]

This split is often portrayed as reflecting the system of dual patronage between those who favored government support and those who wished for more traditional forms of private funding. One might identify Roberts as belonging to a third group existing at this time—one who thought it best to have both sources fund his work so he did not have to depend solely on either. Roberts viewed this arrangement as giving him maximum flexibility while reducing the chances of the total cessation of funding for his work at the HAO. For Roberts, this approach combined both practical and ideological elements—a more robust funding environment giving him more ability to follow his research wherever it led.

The divergence among astronomers over funding, in turn, is often subsumed under wider discussions about the controversies over the very nature of science funding in the post-WWII years. This controversy pitted those who supported Vannevar Bush's more centralized and elitist notions of science against U.S. Senator Harley Kilgore (D-WV). The latter proposed a National Science Foundation that, according to proponents, had a more democratic and less northeast-centric approach to funding science.[90] This controversy was not always a friendly disagreement among the scientists involved. Shapley once referred to Bush supporters as "kept men" of industry.[91] Despite this, the resulting legislation created in 1950 did favor Bush's views. This episode in U.S. science policy may appear as a "victory for political elitism," even though the resulting NSF was far less powerful in setting national science priorities than Bush had envisioned.[92] Bush and Shapley in the end, however, did agree on one thing—the appointment of Alan T. Waterman as NSF's first director.[93]

As indicated in the discussion over the creation of the Sacramento Peak Observatory, the debate over a new NSF was itself imbedded in the controversy over military support for U.S. science. The interactions among Roberts, Menzel, and Shapley in this and subsequent years demonstrated to varying degrees the tensions that arose even among close colleagues with similar political ideologies. The interactions also demonstrate that this was a complex time, with disparate views among even just these three scientists that are not easily described or categorized. These tensions only increased as the fears produced by anti-communist fervor in the nation also arose, producing a generalized "age of anxiety" among many scientists.[94]

According to Roberts, Menzel wanted HAO completely funded by the government as part of the National Bureau of Standards or to become an Air Force operation. This claim that Menzel wanted HAO subsumed under the NBS (via the Central Radio Lab) fits into Menzel's general approach to get the government, military or otherwise, to fund his burgeoning empire consisting of the western solar stations.[95]

The resulting configuration of a private HAO funded in large part by the U.S. military and a government-owned Sacramento Peak was, in Roberts mind, an adequate compromise between these two views—all or nothing—about government and military funding.[96] Aside from its importance to events in Boulder, the episode surrounding the creation of Sacramento Peak showed the complex, if not contradictory, views in the scientific community about accepting government funding, and any association with military sponsors in particular.

Many historical analyses of this era picture Shapley, for example, as being resistant, if not hostile, to military funding of any type.[97] Yet, in his letter to Roberts on arguments for and against the Sacramento Peak project, Shapley sought the "easy" money (Shapley's quotation) despite his well-known views that military money was "blood money."[98] Military money was one of his arguments in favor of the new site, as he wrote to Roberts.

His oft-quoted statements about his revulsion to military funding now appear often rhetorical in purpose, perhaps expressing some ideal state of affairs, for he never actively impeded his subordinates who aggressively pursued such funds, including Menzel, Whipple, and Roberts. Perhaps he simply wanted to have others obtain outside funds for their work so he could better control HCO funds to suit his own purposes.

Further, Shapley's suggestion to close HAO and the Climax site could have put HCO's western activities fully into military hands, with near-total Air Force sponsorship and oversight. Shapley helped advance this proposal without any serious reservations. His practical sense clearly superseded any ethical dilemmas about HCO accepting "blood money." He understood the funding was a "battle for survival."[99]

Shapley at this time also strongly advocated for NSF legislation that freed it from any form of political control, for, as he wrote Roberts, he did fear somewhat "getting under Army control" on a permanent basis.[100] But he also feared any control outside of the scientific establishment, military or otherwise.[101] Shapley served as one of the leaders of the move for an NSF controlled by scientists and helped to form, with chemist Harold C. Urey, a committee for a National Science Foundation to advance this goal.[102] Nominally, this version was the Kilgore bill's version of an NSF, but one wonders if Shapley really supported the populist approach inherent in Kilgore's views.

Given the Sacramento Peak episode, one might more accurately and fairly describe Shapley's views on science funding at the time as realistic, and simply as much preferring civil to military sponsorship. Recalling his deep involvement with the creation of Sacramento Peak, this research demonstrates that claims that Shapley "especially resisted" such funding in the postwar era are a bit overstated and apply only to his own scientific activities, not HCO's.[103] Although he personally may not have sought out or even encouraged military funding of HCO activities and wished for new government sources, he clearly did not force his underlings to shun military sponsorship of these ambitious projects of great importance to HCO staff when they did arise. He understood, after all, as he wrote Roberts, that it was easily available money. These funds, in a practical way, enhanced the work of the HCO even if the money did not support directly his interests in galactic astronomy.[104]

University of Colorado president Stearns also had reservations about military funding and supported the Kilgore proposal. Despite his close associations with the military, the politically aware college president strongly advocated for the creation of a National Science Foundation to supplant the heavy military funding of science. In this he joined most other scientists in the nation, including Harlow Shapley. Stearns stated, for example, in a letter to U.S. Republican Senator Eugene Millikin from Colorado, that he was "a little fearful of the consequences" should the U.S. scientific research program be left "entirely in the hands of military men," the apparent trend at the time. He expressed the desire that scientific research should "be in the hands of people who are broadly concerned with the widening of the human horizon without particular regard to military application."[105]

Even though some scholars have claimed that Roberts fundamentally agreed with Shapley's aversion to military sponsorship, Roberts's views on government and military funding actually fell between his mentors' opinions.[106] Roberts did want HAO to continue as a private organization so as to not get "too close" to the military. However, much, if not most, of Roberts's funding until the late 1950s came from the Air Force or the Navy, and he never expressed aversion to this sponsorship for his research—rather, he eagerly sought funding from any source as long it did not involve classified work.[107] For him, the pure symbolism of a private organization for science was paramount, but his ideas evolved over time in this regard, as later events demonstrate. Roberts, it is important to note, did not appear primarily concerned with vital issues such as losing control of his research agenda because of military funding.[108] Roberts sought something of a middle ground on science funding—he was happy to exploit the entire spectrum of what one scholar called the "rapid proliferation of patrons" for astronomy in this era.[109]

Roberts more often expressed concern (and frustration) over dealing with the involved and sometimes capricious bureaucracy that came with military sponsorship—"the red tape."[110] His work had no classified dimensions and he never expressed, either publicly or privately, any sort of strong ideological distaste for military support. His comments on "red tape" could only have meant he found the military budgetary process onerous. He never appeared to express the same reservations about military control or science in general as did some other physicists of the time, such as Merle Tuve of the Carnegie Institution.[111] In contrast, years later Roberts often expressed appreciation rather than deep ethical concern when speaking of military support.[112] His combination of government (mostly military) and private sponsorship worked well enough for the better part of the 1950s.

Roberts's views in this regard seem echoed by a former Harvard colleague, noted astronomer Leo Goldberg, who considered government contracts to be frosting on the cake, "but never the 'whole cake.'"[113] Roberts, as far as his scientific work went, understood that he had nothing about which to complain, short of the Byzantine bureaucracy and uncertain funding cycles of the modern American military.

Menzel, in contrast, wanted as much government sponsorship in the post-WWII era, civil or military, as he could get to support astrophysical and geophysical studies—to fund the "whole cake," to use Goldberg's analogy.

No doubt worn down by years of continuously searching for private funding for his (and Roberts's) research as Harvard's "solar empire" expanded, the advantages associated with government funding far outweighed potential problems for Menzel. His work in WWII, both as a civilian and then as a naval officer, made him aware of the possibilities of continued government funding for science. His experiences also convinced him of the efficacy of this support.[114] Like his Harvard colleague Fred Whipple, Menzel had indeed found that his prewar work now had "new purpose and direction."[115] He made himself continually alert to these possibilities after 1945.

By the late 1940s, Menzel and Roberts had developed a clear rift in their attitudes on how best to fund their science, a rift that mirrored similar tensions at the HCO in Cambridge and within the broader U.S. astronomy community.[116] This divide only grew greater in subsequent years as other tensions arose relating to the best way to govern the new HAO, and the specter of McCarthy-era loyalty hearings for both men exacerbated these tensions even further.

As Menzel worked closely with the Interservice Radio Propagation Lab, the Central Radio Propagation Laboratory's (CRPL) organizational predecessor, during the war years when he was on active duty with the U.S. Navy, he understood that HAO had natural and potentially permanent government sponsors in both CRPL and the military services. His quest for continued government postwar funding for his solar work by these sources presented him (and his protégé Roberts) with a rare opportunity—the creation of the first major element of sun–earth science in Colorado with HAO. In this, HAO became part of Menzel's "rapidly expanding solar empire."[117]

Roberts, first by creating HAO and then by helping to fend off threats to the nascent solar observatory operations in Boulder, ensured solar physics stayed in Colorado, and thus began the process of turning Boulder into a very different kind of place on the map of the United States—a city of science. The first major change would be CRPL's relocation to Boulder from Washington, D.C. The next chapter covers the complex story of how scientists, policy-makers, government officials, and local citizens worked to make this the move possible.

CHAPTER FOUR

A Scientific Peak Begins to Develop in Boulder

In the late 1940s and early 1950s, U.S. government sponsorship of earth sciences increased to unprecedented levels.[1] As a result, federal patronage related to the sun–earth connection enabled scientific activities in addition to HAO to begin in Boulder. Scholars have emphasized the role of military funding in science endeavors at this time, but what makes Boulder particularly interesting in this regard is that the next major funding of sun–earth science came from a civil source: the National Bureau of Standards (NBS). Scholars argue that NBS funding was military funding in civilian garb, but this underestimates the need of society, and not just the military, to better understand radio propagation and associated phenomena.[2] Roberts and Menzel, the Boulder Chamber of Commerce, the wider local community, elected officials, and policy-makers all played significant roles in using the opportunities of the era to increase the amount of sun–earth science in the city. This chapter analyzes how the relocation of an important science facility to Boulder happened.

Expanding Sun–Earth Science in Boulder

Roberts and Menzel, ever the scientific entrepreneurs, did not only concern themselves with the activities of HAO alone; they also tried at this time to build a wider sun–earth connection scientific community in Boulder. They did this to create a more stimulating milieu for HAO's work and to decrease the sense of scientific isolation that existed for Roberts (and other HAO staff). Roberts, already committed to his new hometown, had a desire to help Boulder grow and prosper in the postwar years as well.

Menzel and Roberts started to create, albeit unknowingly, what some refer to as a "Marshallian" district. Adapting the work of economist Alfred Marshall, Christophe Lécuyer explained the creation of Silicon Valley as a type of Marshallian district.[3] In these regions, skilled workers employ specialized manufacturing knowledge to create new centers of industry. In the process, these centers attract more workers with subsidiary skills, all the while increasing local economic advantage. In Boulder, the specialized knowledge was sun–earth science. The subsidiary workers were the equipment builders, electronics technicians, and the like. Boulder's Marshallian district, then, is one centered on the production of scientific knowledge, not industry.

Some of this early sun–earth research in the city in the immediate postwar years began on a modest scale, but this does not negate its importance to the development of Boulder as a city of scientific knowledge production. The University of Colorado benefited directly from the Air Force's new interest in sun–earth connection research. Marcus O'Day, an Air Force Cambridge Research Center scientist, as part of his research effort in solar physics, needed a device to stabilize a rocket-borne coronagraph.[4] He turned to the University of Colorado for help with this difficult problem.

The "pointing control," the subsequent name of the system, was a complicated instrumentation problem given the varying movements of an ascending V-2 rocket. The work on such a sophisticated device perfectly complimented the research underway at Climax and planned for Sacramento Peak. Menzel (and Roberts in Boulder) quickly recognized another opportunity to assist their own efforts by enhancing the understanding of how the sun affects the earth's upper atmosphere and by creating another element in Boulder's growing scientific community. They therefore pointed O'Day to the small, but ambitious, University of Colorado physics department.[5]

The physics department chair, William B. Pietenpol, accepted the difficult challenge of creating this pointing control system for the V-2's coronagraph. He created a group under the university's Engineering Experiment Station to support the project. Work started in April 1948, and by early 1949 the Upper Air Laboratory (UAL) on the Boulder campus evolved from the effort.[6] Roberts and Evans served as advisors to the UAL, thereby further solidifying both Roberts's and HAO's association with the school. The $69,000 Air Force contract to build the pointing controls for the rocket coronagraph represented a modest, but significant, start for space and atmospheric science research in Boulder.[7] As important as these efforts were in the growth of the University of Colorado's development as a site for modern scientific research, bigger opportunities for space and atmospheric science soon presented themselves: HAO and Walter Roberts would provide an important link in the chain of circumstances that in 1954 brought to Boulder its first major national research facility, the Central Radio Propagation Laboratory (CRPL) of the U.S. Department of Commerce's National Bureau of Standards.

Bringing Science to "Scientific Siberia"

U.S. government officials first thought about moving the CRPL facility around 1948. CRPL was the nation's center for predicting radio propagation, and as such, stood at the center of sun–earth connection research. It also was an important customer for HAO solar data and a primary source of funds for work. By working to get the lab relocated from Washington, D.C., to a then very remote Boulder, Roberts and others involved laid the foundation for transforming Boulder from a scientific backwater to a scientific powerhouse. The importance of the lab's relocation was not merely in the number of scientists and other workers moved to the city—the effect was more profound. It significantly increased Boulder's reputation in the 1950s as a place to do sun–earth science and therefore made it one of the research hubs of the International Geophysical Year. Harkening back to Wickliffe Rose's sentiment, Roberts and his associates did not just "make a peak higher," they actually began to build the "peak" from almost nothing by creating HAO and getting CRPL relocated to the city.[8] This peak was built upon the foundation of a Marshallian district for scientific knowledge production.

Ionospheric physicist Alan Shapley, Harlow's son, related that in the 1940s the northeastern-based scientific establishment viewed Boulder as a "scientific Siberia," because it was "west of the Hudson River."[9] However, Chicago, Berkley, and Pasadena were hardly "Siberias" for science. In his hyperbole, Shapley likely reflected the scientific establishment's view of the time that only existing centers of learning, mostly in the northeast, could produce the best science, and therefore it was a waste of resources to try to encourage the growth of other locations for scientific knowledge production. This elitist attitude stood at the center of Rose's idea to "make the peaks higher."[10] These "peaks" were places, and not just institutions, of scientific research that many viewed as elite.[11] Senator Harley Kilgore's version of a National Science

Foundation, ultimately defeated in congress, was his attempt to overcome this elitist attitude via the political process. He hoped to do this by spreading science around the nation in part through geographic distribution of U.S. research funding.[12]

A number of factors converged for Boulder to become the location of the new CRPL facility despite the scientific establishment's attitudes. Contrary to popular wisdom, these factors do not include that Mamie Eisenhower hailed from Colorado, or that the move resulted from a formal plan to disperse important government facilities around the United States.[13] It was, rather, a combination of general early Cold War defense polices and, perhaps more importantly, political machination that brought the CRPL to the west in the foothills of the Rockies.

Most renditions of the NBS move to Boulder begin with the account of National Bureau of Standards director, Edward U. Condon.[14] Noted physicist Condon, head of the NBS from 1945 to 1951, wrote in a letter to his wife in the summer of 1949 that the government wanted to build the new lab facility outside of the Washington, D.C., area for generic "military strategic considerations."[15] Condon composed the letter summarizing his thoughts after Menzel and Roberts invited him to spend time touring Colorado following a scientific conference on cosmic rays in Idaho Springs.[16]

In the letter, he told his wife what he was "scheming." "I had been thinking of Princeton," he wrote. He had changed his mind and thought "the best possibility is to put it on the campus . . . here in Boulder." The city had, he wrote, a good "central location" in the United States, nearness to the (Climax) solar observatory with which, as noted, CRPL had an intimate working relationship, and "close relation to a good university." He also thought the area would be a good "summer capitol [*sic*]" for the head of NBS. Condon thought he could combine work with "beautiful mountain vacations." He ended his missive to his wife with the final thought that "if things go well we will get Boulder established as a major branch of NBS and spend our summers out here regularly."[17] Most works recounting CRPL's relocation to Boulder take this letter on face value, especially as it is a contemporaneous document and is consistent with accounts given by Condon.[18]

This letter, and subsequent statements over the years made by Condon, indicate that Condon thought of the idea to relocate to Boulder and was the primary force behind the relocation. The documentary evidence suggests, on the other hand, at least one other version of the genesis of the idea for CRPL's relocation to Colorado. In this alternate version, others actually forced Condon to make the move west, and the true motive for the long-distance move in reality had little to do with national security concerns of the time. The prime mover in this alternate understanding is powerful Colorado Senator Edwin C. "Big Ed" Johnson.

Researchers to date seem to have underappreciated the senator's true role in these events: in their emphasis on investigating scientists and their military patrons, scholars often miss the role of the political process in the development of modern U.S. science.[19] Government agencies, even the military at the height of the Cold War, rarely received blank checks from the Congress. Events in Boulder and Johnson's role in the CRPL relocation clearly demonstrate this fact of American political life.

"It has been rumored that Dr. Condon brought the Central Radio Propagation Laboratory to Boulder," wrote retired Senator Johnson to Francis W. "Franny" Reich of the Boulder Chamber of Commerce in 1967. He added, "The truth is that this tremendously famous Laboratory brought Dr. Condon to Boulder."[20]

No doubt Johnson referred to Condon's claim, often made after he moved back to Boulder in the 1960s, that he started the process that brought the lab to the city in 1949. Most accounts relate that Johnson did play an important part in the NBS move, but only after Condon (inspired by Roberts and Menzel) conceived of the idea. The definitive history of the CRPL, while acknowledging Johnson in a footnote, mentions the Colorado senator only in passing as a helpful politician who had played a "major role" in bringing the CRPL to Boulder.[21] Often, scholars associate Johnson's activities with the period after Condon supposedly decided upon Boulder.

A revised account places Johnson at the center of activities. This politician, elected Colorado governor and senator three times each, remained a popular and influential politician throughout his career, despite being a Democrat in a mostly Republican state. This popularity resulted in large part from Johnson's demonstrated ability to bring government projects to his developing western state. Not the least of these projects was the Rocky Flats nuclear weapons plant and the routing of a major national road, Interstate 70, through Colorado.[22] By having key positions on the Military Affairs Committee and the Joint Committee on Atomic Energy in the post-WWII era, he brought modern sun–earth research, in addition to many other activities, to the state.[23]

The CRPL Move

The complex sequence of events that brought CRPL to Boulder began on May 26, 1948. U.S. Secretary of Commerce, Charles Sawyer, wrote to Senator

Wallace H. White, Jr. (D-Maine), chairman of the Committee on Interstate and Foreign Commerce, about a proposed bill, S. 2613, to create a new CRPL facility. In addition to submitting justification for the new lab, Sawyer added that although not specified in the legislation, he "contemplated" the new facility "will be erected within the grounds of the Bureau of Standards in the District of Columbia."[24] This bill soon died from lack of support, but this did not end the process.

Soon after the Democrats won control of the Senate, in January 1949 Johnson became chair of the Senate Committee on Commerce. According to Johnson, Condon's assistant, Hugh Odishaw, came to discuss with him reintroduction of the NBS lab bill. Odishaw expressed a strong desire, apparently reflecting Condon's thinking, to keep the new lab within 100 miles of Washington, D.C., even given the general desire of the government to disperse federal facilities away from the nation's capital. Johnson recounts telling Odishaw, "When you are ready to agree to not interfere with Congress on the location" of the lab, "come back and I will introduce your bill . . . I am strong for the construction of this badly needed laboratory."[25] Edward Cooper, an aide to Johnson involved in these discussions, later wrote that "I think it did not take Mr. Odishaw too long to understand that the only chance for enactment of the legislation was if the Bureau did some hard thinking about another location."[26]

Johnson claimed that Odishaw returned in two weeks with most of the "assurances" he had "demanded" of the NBS. He added that "Odishaw had some pertinent questions about the cost of Boulder land."[27] If this account is accurate, this conversation took place somewhere in mid-January 1949. This date generally coincides with Johnson's resubmission of the bill on January 13. It is also months before Condon's trip to Colorado.

The new bill, S. 443, did not include provisions for a site, as that decision remained within the purview of Secretary Sawyer. On February 24, Sawyer sent Johnson a statement of justification for the bill. The bill still mentioned the grounds of the NBS in D.C. as the probable site. It left open a possible relocation "as a security measure in the event of a national emergency."[28] In a subsequent letter to Johnson on June 2, Sawyer informed the Senator that he "understands that concern has been expressed by several members of your committee" about the CRPL site and that "I am now of the opinion that this building should not be located" in the environs of D.C. This sudden shift of position occurred in part, the Secretary claimed, due to his consultation with members of the "National Military Establishment."[29] These events opened the road to Boulder. As Johnson's aide Cooper put it, "if Senator Johnson had not taken the ball and run with it, the radio propagation laboratory, if it was built at all, would be located on the Bureau's grounds within the District of Columbia."[30]

If Johnson and his former aide's rendition of events is even generally accurate, it appears that Condon arrived at the cosmic ray conference at Echo Lake in late June 1949 already thinking of Colorado as a potential, if not probable, site for CRPL. Senator Johnson had made sure of that. Condon's later claims that he conceived of the move at the time of his Colorado visit become problematic given this evidence. By June, the process of relocating CRPL to Boulder began in earnest.

Condon, prompted by Menzel and Roberts, began his Colorado sojourn with visits to Boulder, Denver, and Climax, followed by a short fishing trip. In the letter to his wife, Condon wrote that he was "too lazy to do any fishing," but that he took some long hikes in "some of the most beautiful country I have ever seen." The group returned to Boulder for a dinner party at Roberts's house that included four recently arrived officials from the NBS. Some of these men soon served on the site selection board for the NBS lab. Once he returned to Washington, Condon added, "I am going to plunge into the Washington end of the job of establishing our new radio labs out here."[31] Roberts and the others could not have known at the time that Senator Johnson appears to have made their task of relocating CRPL much easier.

Condon visited Governor John Knous, "A Democrat and a good friend of Truman and also of Senator Johnson."[32] The letter to Mrs. Condon gave no inkling that Johnson and the NBS already had extensive interactions about the CRPL move prior to this meeting with Colorado's governor. Nor did the missive event hint at Johnson's early interventions on behalf of the relocation.

Condon also visited President Stearns of the University of Colorado. According to Condon, both Knous and Stearns were "very enthusiastic" about getting the new facility in Colorado, presumably in Boulder near the university. As Condon wrote his wife, this overwhelmingly positive reception resulted in his summoning the aforementioned NBS staff. Together they would look at the "possibilities" of getting the lab established in Boulder. One of the staff he sent for was Alan Shapley, Harlow's son and Roberts's long-time associate.

As Roberts later related, "We took them up to the top of Flagstaff Mountain on a day when the temperature in Washington was about a hundred and the humidity ninety-nine."[33] The NBS group then returned to Washington with "the boys," that is, the staff Condon had sent for previously, "all enthusiastic" about a Boulder site for the new NBS center.[34]

Even with what we now know to be Condon's (prompted by Johnson's) strong sentiments on this matter, this major $4.5 million effort by the U.S. government needed strong justification to the public, and, in particular, other politicians. Officials therefore had to create a formal and impartial site selection process. Officials formed a four-person site selection committee that included three of the four staff whom Condon sent during his visit to Colorado. Although Alan Shapley was not an official member of the board, he served as an unofficial board secretary of sorts. He was in continuous communication on the move with Roberts, ostensibly because of the professional and scientific link between HAO and CRPL.

According to Alan Shapley, the delegation of NBS personnel that Condon ordered to Colorado in July 1949—a month before the official announcement that a process would soon begin to find a suitable location for CRPL—agreed they would recommend Boulder as the site for CRPL on the flight west after Condon summoned them.[35] The subsequent site selection process began with a strong favorite for the CRPL facility. In fact, the result seemed in little doubt. "If the truth be known, our minds were already made up before we reached Boulder," Alan Shapley mentioned in an interview years later. He added, "When McNish, Smith, Welch, and Norton saw Boulder they could find almost nothing negative to say."[36]

Despite this strong desire by Condon and other NBS officials to locate the new laboratory in Boulder, the formal decision process allowed for at least some ambiguity in the final outcome. Irrespective of the seemingly preordained result that Shapley recalls (and that was desired by Senator Johnson), officials still needed to build a strong case for their decision, for they might have to defend it to the public and politicians who had an interest in obtaining the site for a constituent location.

As other cities became aware of the NBS move and offered their own proposals for possible sites, the NBS personnel and Secretary Sawyer needed help in justifying a Boulder selection. This gave Condon firm rationale to support the decision he had made, apparently under pressure from "Big Ed," long before the formal selection process began. This crucial help came from the citizens of Boulder.

Local Forces Mobilize

Roberts had an important role in bringing sun–earth science to Boulder, as did various governmental officials. In this process of site selection for the NBS lab, the role of a third important element appears—the local community. Boulder, because of its small size and relatively homogenous population in this period, presents an interesting case study since much of the local population played direct roles in bringing science to the city. Unlike regional development in other areas when development was mostly driven by high-level political and bureaucratic elites, in Boulder citizens "voted" for science and thus helped to create the scientific version of a Marshallian district. They did this by contributing money to buy land with the purpose of enticing the government to locate the new lab in their city. This is important to note—unlike other science places or regions where state and local officials ceded public lands to bring a scientific or military site to a community, here the citizens banded together to obtain private land so they could give it to the government.[37]

Although not used today, the practice of offering land as an enticement for the award of a federal facility was not uncommon in the 1940s through the 1960s. The chamber of commerce in Colorado Springs used such a ploy to obtain military bases—Camp Carson and Peterson Air Force Base—in both WWII and the Korean War.[38] Even earlier, the Ogden, Utah, chamber of commerce offered land to the government for what is today Hill Air Force Base.[39] Later in the era, both the U.S. Air Force Academy in Colorado Springs and Houston's NASA Manned Spaceflight Center (now Johnson Space Center) went to locations that had donated either public or private land to the government.[40] Similarly, San Diego city government offered choice land in La Jolla to bring the private Salk Institute there in the early 1960s.[41]

In the Boulder story, however, the Chamber of Commerce obtained the funds to buy the land in an actual fund drive, a drive where many locals contributed. This was not just a case where a number of businesses or leading businessmen pooled resources. The added twist on the Boulder story is that it was perhaps the first and only time citizens banded together to bring scientific laboratory, rather than a military or industrial facility, to their location.

Events proceeded quickly on the lab relocation after Condon's summer sojourn to Colorado. As reported in Boulder's local newspaper, *The Daily Camera*, Senator Ed Johnson pushed for final action on the congressional bill that both authorized the new facility and empowered NBS director Condon to select the location.

The existing Air Force-sponsored activities at the University of Colorado and the mostly ONR–Air Force funding of HAO paled in comparison to the prospect of the development of a major, state-of-the-art, multi-million-dollar federal government research center in the heart of Boulder. Roberts, Stearns, and the community rallied around the idea, for disparate reasons, of making this enticing prospect a reality.[42]

Local elected officials on the city council were not active in leading the charge and energizing the community to get the NBS facility to Boulder, and why this is so remains unexplained. Rather, it was the Boulder Chamber of Commerce that immediately seized upon the opportunity to exploit this possibility of largess from the federal government. Chamber of Commerce president John Allardice and secretary Franny Reich, "are in touch with Colorado's members of Congress," reported a Boulder paper.[43] The Chamber planned the campaign not just as a local effort, but extended it to the halls and lobbies of Congress as well.

Allardice and Reich communicated with others beyond Colorado's elected officials on the NBS move, including Walter Roberts. For the next few months, as the campaign intensified, Roberts served as an important link connecting the Bureau's site selection committee and the Boulder Chamber of Commerce. His unique position in the world of the sun–earth connection and high interest in the establishment of the new lab made him ideally suited to serve in this role.

Roberts corresponded with Allardice on August 24 about the "location of NBS electronics laboratory in Boulder." He wrote that he was "speaking entirely for myself, and without instruction, or even knowledge of, any persons in the government." Roberts, however, had been involved in courting Condon and his staff since at least the June Echo Lake cosmic ray conference. Roberts went on to list for Allardice what "questions will enter into consideration by the director of the NBS as he studies possible future locations for the laboratories." Among the topics he listed were scientific suitability, university, community, and "protection against speculative buying of land and housing." "Dr. Condon," Roberts added, "will make a fully impartial evaluation of all sites which have special merit." Therefore, "it will be important for us to begin to assembly [sic] well-formulated information about the special merits of Boulder." Roberts added that tax information and water supply might be among the issues of concern in this August 24 letter.[44]

"One thing we didn't get quite straight when we were to see you in early July," Alan Shapley wrote to Roberts only a day later, remained "namely the real estate tax situation in the city and county." Could Roberts, Shapley asked, provide information on other taxes? "Some of us (not me in particular) have been worrying about things like this and would appreciate more dope."[45] Roberts became a pipeline of information between the NBS and the Chamber, an ideal arrangement for all involved. Roberts was very much in tune with Condon, Shapley, and others' thinking about the NBS move.

Roberts responded dutifully on the tax question to Shapley, and also asked if it was possible to get an estimate of the number of NBS employees who might move to Boulder. This information could help Roberts find out if "housing can be provided at firmly guaranteed prices, when we reach a more definite state of negotiation."[46] In this correspondence, ostensibly simply about information exchange and well before the formal site selection process ended, Roberts laid the foundation for the move to Boulder.

The Selection Process Begins

The NBS did create a formal process of site selection, a process Boulder advocates needed to navigate. Condon selected Newbern Smith, A. G. McNish, K. A. Norton, and S. W. Welch for the formal site selection board. The board established seven criteria for site selection: the site should be in a small city or town, near an "adequate" university (defined as offering a PhD in physics and merely having an electrical engineering department), near a large city, located in a "non-congested area," have a moderate climate, diverse terrain, and be easily accessible. These criteria ensured that Boulder would be a front-runner in the competition regardless of what other cities might also make it to the top of the selection list.

Using the first two criteria, twenty-eight schools in towns of less than 100,000 people made the first cut. The board reduced this list to seven contenders by applying the other criteria, and another round reduced the original list of fifty-three to three: Palo Alto, California; Charlottesville, Virginia; and Boulder. The board considered all three as "suitable" candidates for the new lab.[47]

The board needed a final distinguishing factor, and one that supported Condon, Shapley, Roberts, and (perhaps most importantly) Senator Johnson's desire for the Boulder location. They found the needed differentiator in a land offer to the U.S. government, for this offer could tip the balance toward justifying the choice that they all wanted in the first place. Roberts played a key role in this negotiation.

Roberts wrote to Shapley on October 7 that the Boulder Chamber of Commerce was "unusually active and progressive," and Boulder "is prepared to take specific and definitive actions to assist and support anything like the Bureau move." The Chamber had already undertaken optioning choice land tracts in Boulder for a period of nine months. "We felt," wrote Roberts in his letter, "that it was necessary for them to do this to protect the Bureau against any possible rise in land values" resulting from a newer road to Denver and speculation about "the possibilities of the Bureau's move" once that news became generally known.[48]

In a coincidence of timing, on that very day, Alan Shapley wrote to Roberts about the same topic of land. "The way these things work," he commented, "the *free availability* of land may become an important factor in the selection of a site." Roberts, in turn, responded. He had talked to Reich by this time about the land issue, in part to prepare him for an upcoming trip to Washington. Initially, the "reaction of the business people" in Boulder was hesitant; they did not "wish to make any definite decisions or promises" about the land issue. He recommended the Chamber of Commerce should take up the issue as "they probably would do something."[49] Of course, the Boulderites had no way of knowing that the biggest factor in their favor was the desire of the major players in this story to have the site in Boulder in the first place. The land issue became a lever for Condon and the rest to translate their desire into reality.

Two days before Roberts wrote this letter, he had a phone conversation with Menzel. They discussed the NBS move, in addition to HAO matters. Roberts told Menzel that the Chamber of Commerce was "worried" about the site selection process. Menzel then elected to call Condon himself "tonight or tomorrow" to see "if anything should be done here."[50]

Menzel soon reported to Roberts the substance of his talk with Condon. Condon in effect reassured the citizens of Boulder, through Menzel and Roberts, that things were "OK so far as Congress is concerned." That is, all remained on track for the bill to pass authorizing the move. He added that they should "keep in touch with Ed Johnson."[51]

In addressing the "donation of land problem," Roberts wrote that although the land was not a "major expense," "we all appreciate here 'the political angle.'"[52] However, as the selection process continued, the free land issue rapidly increased in importance as a factor in getting the NBS lab to Boulder. It became clear that there were viable competitors for the site.

Reich returned from his trip to D.C. where he showed promotional films about Boulder and the University to Condon and others at NBS. The Chamber of Commerce voted on October 25 to "leave no stone unturned to secure the laboratory," and one of the stones unturned was to offer land to the government for the site "if necessary."[53] Although it is uncertain what transpired to transform the possibility of donating land to a guarantee in this period, it did soon happen.

The Boulder supporters of the move went into high gear once Congress finally passed the new lab's authorization bill on October 19. Only a week after Reich's trip, on November 2, the chamber and the university launched their effort to acquire the lab.

The Chamber of Commerce planned to purchase and donate the land, a privately owned parcel of 217 acres worth about $70,000, to the U.S. government on behalf of the community.[54] This information quickly made its way to both the federal government and Colorado representatives in Congress, and "Big Ed," when contacted, once again expressed enthusiasm for the lab and an optimistic outcome for Boulder.[55] Even Denver mayor Quigg Newton, later to become the University of Colorado president, weighed in and asked Commerce Secretary Sawyer, Condon's boss, to look favorably on a Boulder decision.[56]

The land offer proved essential to the board's final selection. The offer made it easier for them to justify, package, and sell the decision that they and Condon ostensibly made months before in the mountains of Colorado. As it turned out, based on the top three contenders—Palo Alto, Charlottesville, and Boulder—all stood remarkably close in the final tally. Although Boulder achieved first place, only two points separated Boulder from its two top rivals. Given a rating from one to three on each of the selection criteria, the final score was Boulder thirteen, Charlottesville twelve, and Palo Alto eleven. Despite Condon's and many others' stated preference for the city at the lab's new home, the contest wound up still close enough for other representatives of the other locations to contest the selection. That there was no clear winner for this important decision held obvious political ramifications and possible controversy. What was the selection board to do?[57]

This is where the Boulder Chamber of Commerce's land offer yielded immediate returns. "Of the three primary locations under consideration," wrote the NBS board in this final report to Condon on December 13, "only one, Boulder, has formally offered a suitable tract of land, comprising 210 acres, to the Federal Government." They added, as if to accentuate the point, that no other sites offered free land "although key officials at the University of Virginia and at Stanford University are aware of the projected relocation of the Laboratory." The report also commented on the extensive study completed of the Boulder location. "In view of the above considerations it is recommended," they wrote, "that the Laboratory be located at Boulder . . . and that the tract of land offered . . . be accepted."[58] In December, on the basis of the NBS report, the Secretary of Commerce made the decision to move to Colorado official policy.[59] Now the Boulder community had to raise the money for their land pledge.

"Buy Prosperity Insurance!"

Commerce Secretary Sawyer, no doubt to the joy of "Big Ed" Johnson and other members of Colorado's congressional delegation, approved the NBS/ CRPL committee recommendation. The good news elated the Boulder community. The *Daily Camera* had served as an advocate of the move and published a number of stories around the time of the official announcement. The day before, it ran a feature article, "Boulder Expected to Be Picked as Site of $4,5000,000 Lab" where it anticipated the announcement on the following day. The article outlined the role of the new lab and its benefits to the community. The paper, as it "had been closely associating with the committee working on the project," could not say anything until the NBS made an official announcement.[60]

The Chamber of Commerce, having achieved this victory in getting Boulder selected as the site for the new lab, now had to deliver the land to the government. This required collecting the estimated purchase price of $70,000 for the land pledged to the government. Where would they find the funds?

"Boulder Citizens to Be Asked for Nearly $70,000 to Purchase Laboratory Site" read a subheading to a February 27, 1950, *Daily Camera* article about the Chamber of Commerce's campaign to raise funds. The Chamber decided on April 11, 1950, as the kickoff day to begin the drive to solicit funds directly from Boulder's citizenry. However, efforts entered a full-swing phase well before then as the proponents of the plan began to marshal the resources of the entire community.[61]

James Yeager, a local businessman, headed the campaign to raise the funds and secure land for the new lab. A key member of the Chamber of Commerce's campaign team was A. A. Paddock, the owner–publisher of the *Daily Camera*. The paper threw its support fully behind the effort to engage the Boulder community in the fund drive.[62]

Both Condon and Menzel arrived in Boulder in early March for a well-timed visit. "The mere fact that Dr. [Condon] is here shows the project is making progress," commented Franny Reich.[63] This visit served as something of a kickoff for the local fundraising campaign to bring the lab to Boulder. Soon after, the seemingly removed Boulder city council finally made itself heard; on March 22 they voted unanimously to simply approve the Chamber's efforts.[64] It remains unknown why the council took such a passive role in these events. Perhaps they showed a reluctance to get involved as the work of Senator Johnson, the Chamber of Commerce, and Roberts proved so effective.

Yeager soon picked ten "campaign captains" to coordinate efforts, but the university, as a state entity, acted separately.[65] The *Daily Camera* ran articles with titles such as "Entire Community to Benefit through Bureau of Standards" and "Engineering College to Benefit from Radio Laboratory." An editorial in the paper had the title "Bureau of Standards Laboratory Brings City Publicity that Is Not 'For Sale.'" It chastised "those who criticized giving a site to a government that spends billions recklessly" for not appreciating what the lab would eventually bring to the community.[66] A small minority of citizens expressed irritation that the city had to give land to the federal government in order to obtain the NBS lab and argued that the government had enough money to buy land if they wanted it. No one appears to have voiced concerns over community growth issues.

The Rotary Club and local unions through the Boulder Trades Council stood among the first to make pledges, indicating the diversity of support for the drive.[67] The Adolph Coors Company and the local Elks Lodge also contributed, and prominent citizens with diverse interests participated as well. The principal of Boulder's high school and the head of the local American Legion wrote statements of support published in the *Camera*.[68]

Small groups also helped. The *Camera* proudly reported that ten of its employees, "desiring to help," contributed $36 each to fund one acre of the land buy.[69] The drive had quickly and successfully developed a broad base of community support.

A further boost came when the IRS informed the Chamber of Commerce that donations to the land purchase drive were tax deductible.[70] Although no direct evidence on this exists, it might be fair speculation that "Big Ed" and possibly other Colorado elected officials influenced this decision. In any event, the campaign started promisingly, and broad segments of the Boulder community became quickly involved in the drive to bring science to their city.

A major part of the solicitation effort as constructed by Yeager and his team included the notion that one could think of a donation as a type of "prosperity insurance." "You probably now have health, accident and life insurance," ran a full-page ad in the *Daily Camera*, "but how about your business?" It added, "Will you have the customers and clients next year you have now?" After listing additional questions, it asked, "Don't you wish you knew?" Bringing the lab to Boulder should bring a $2 million per year payroll "dividend" to Boulder for the "premium" of the $70,000 needed for the site. The page spread ended enthusiastically and optimistically with "pay your share of the premium—TODAY!" and "collect your share of the dividends—TOMORROW!"[71]

The Chamber of Commerce also printed faux "insurance policies" for distribution. "For Every $1 Invested in This Campaign," the policy claimed, "The Community will annually get back $28.50 in new payroll."[72] Irrespective of the merits of this advertising approach, the brief campaign succeeded well beyond the originally stated goals. On April 20, the Chamber declared "victory" in this effort. As of that date, more than 600 donors or groups of donors had contributed over $80,000, well exceeding the goal of the drive.

Seeing a successful conclusion to their all-out effort, the Chamber organized a "Victory Breakfast" on that date. It invited "Big Ed" Johnson to attend, as well as the many campaign workers. Yeager and Reich gave the Senator a "promissory note" for the NBS site, yet another faux document, posted on a large piece of cardboard to take back to Washington.[73] "I am confident," the senator said, seeming to echo the fund drive, "Boulder citizens will find the bureau the greatest investment they ever made."[74] The senator, despite his hyperbole, did speak the truth, in a way. A true intellectual community based on the sun–earth connection began in the city—CRPL became a key addition in the city's rapidly developing scientific complex and a major anchor for space and atmospheric science there. A new scientific era came to Boulder with the arrival of this NBS lab as a scientific Marshallian district began to develop.

Originally inspired by the Colorado senator, the nearly year-long process to implement Condon's vision of his "summer capital" succeeded. This success was due to the citizens, and businesses, of Boulder via their cash donations, in effect, to the U.S. government. The "scheming" the NBS director told his wife about in his July letter had, with the help of Johnson, Roberts, and many others in the community, worked in the end.

A New Type of Frontier

The Boulder Chamber of Commerce deeded the land to the U.S. government on June 13, 1950. Construction on the new NBS buildings started soon after. The facility had almost 180,000 square feet of space. As a finishing touch, a plaque honoring those who contributed more than $100 in the Chamber of Commerce drive adorned the new building.[75]

President Eisenhower, the first incumbent U.S. president to ever visit Boulder, dedicated the NBS CRPL building on September 14, 1954. Eisenhower spoke of the labs as representing a "new type of frontier," a frontier "of greater romantic value and greater material value than some of the discoveries of those earlier days."[76] He invoked the metaphor of scientific advance as a new frontier almost a decade before the Kennedy administration used the same trope, as one historian commented, to the point of "cliché" with respect to outer space.[77]

The president's comments on this occasion appear to invoke the thoughts of noted U.S. science policy-maker Vannevar Bush given in his work from 1945, *Science—The Endless Frontier*. Bush's title, in turn, derived from President Franklin Roosevelt's charge to think about the "new frontiers of the mind" that modern science presented.[78] Despite the use of the frontier trope, there is no evidence that Eisenhower or his speechwriters actually had Bush in mind. The president, himself a product of the vanishing American frontier (Abilene, Kansas) and an avid reader of Wild West stories, invoked an idea that simply resonated both with him and with the westerners he addressed on that late summer day in 1954.[79] The idea of science as a frontier extended to a new dimension the importance of frontier in American thought and ethos—an idea proposed by Frederick Jackson Turner in 1893.[80] By relating the expected scientific advances of the new NBS labs to the historic experience of the American West represented by the city of Boulder, President Eisenhower attempted to link the past of both the nation and region to the future of Boulder.

With HAO, CRPL, and the sun–earth science work underway at the University of Colorado, by the early 1950s there existed a firm organizational, institutional, and intellectual foundation that enabled continued growth for Boulder as a world center for sun–earth studies and other sciences as well. Some scholars point to the importance of non-hierarchical configuration of activities in explaining the growth of science regions in the United States, and this is precisely what transpired in Boulder.[81] A horizontally integrated, intellectually bound scientific community based on the compelling and complex questions relating to the sun–earth connection spontaneously evolved in the city, a community that would make further growth possible given the right conditions.

As noted, this creation of non-hierarchal configuration is a component of Marshallian districts. As was the case of Silicon Valley discussed by Saxenian and Lécuyer, the close communication and open structure of scientific discourse that characterized Boulder's science activity enhanced the creation of Boulder's sun–earth research community. To use Lécuyer's term, there existed in Boulder easy and open "access" to all who wished to participate in the sun–earth studies community.[82] Shared scientific interests; dependence on each other for funds, data, or knowledge; and the fact that many of the scientists involved—Roberts, Menzel, Alan Shapley—already had personal and working relationships facilitated and strengthened the growth of Boulder sun–earth science in the 1950s.

As was the case in Silicon Valley, the sun–earth scientific community extended well beyond the immediate city.[83] This enabled Boulder-based scientists such as Roberts to not only extend their influence in the discipline, but to also begin making Boulder widely known as a place for this research. Unlike initial Silicon Valley developments that tended to focus on the San Francisco peninsula, sun–earth science in Boulder had, from the beginning, not just regional but global aspects. For all these reasons, Boulder's development as a city of knowledge production generally accelerated in the 1950s.

Roberts first derailed Harvard's suggestion of withdrawing from HAO and then played a vital part in the CRPL relocation campaign. In doing this, Roberts helped to create the beginning of a scientific "peak" in Boulder by establishing an institutional foundation for a Marshallian district based on sun–earth research. In the decade of the 1950s, he would continue to play this important role in making the nascent science peak even higher. Before this happened, however, Roberts would face a number of significant institutional and funding challenges.

CHAPTER FIVE

"Nothing but a Fundraiser"

The successful bid to get the CRPL in Boulder was an important step in the city's growth as a site for science. Yet significant challenges would face Roberts and Boulder sun–earth science in the early 1950s. Although *Physics Today* declared in 1950 that "the springtime of Big Physics has arrived," this time proved anything but tranquil for Roberts's efforts.[1] Until WWII, private patronage proved essential to the construction of new astronomical facilities.[2] Roberts thought of this type of support as an important augment to the funds the military and the NBS provided for his work. He accepted the new government-provided funds, as did Menzel, but refused to let go of the traditional ways in the style of Harlow Shapley. This role as a transitional figure between the traditional and newer attitudes on science funding had costs for Roberts. It strained his relationship with Menzel and required convincing potential donors about the usefulness of his seemingly esoteric astrophysical studies.

In this chapter we see how the early 1950s became for Roberts a period of intense fundraising from private donors to sponsor his ambitious sun–earth science efforts in Boulder. It also was a time of growing friction with Menzel not only over funding, but also over HAO control and security investigations they both endured. Roberts's idea to use the sun–weather connection (a concept he would later term "astro-geophysics") as his primary fundraising tool and his eventual professional split with Menzel (and HAO's from Harvard) set the stage for Boulder's continued, rapid growth as a city of scientific knowledge production during the early Cold War.

Roberts the Fundraiser

"You're nothing but a fundraiser" irate teenager David Roberts told his father Walter during a heated argument in the mid-1950s over the need for curfews.[3] The adolescent attempted, and apparently succeeded, to strike a painful blow directly at a point of vulnerability in his father's persona—Roberts's emerging role as a scientific entrepreneur and the associated decline in his scientific research efforts. The younger Roberts understood the tension that arose in his father's life between these competing demands. Roberts's attempts to integrate all of these activities shaped not only his career, but also events in Boulder.

By 1949, a year after the return from his Harvard sabbatical, the HAO had developed into a "fairly substantial" enterprise.[4] With work centers at Climax, Boulder, and Sacramento Peak, the HAO staff grew to almost thirty scientific and support personnel.[5] Besides operating the old coronagraph at Climax, Roberts's team busied themselves at Climax and in Boulder workshops with the construction of improved coronagraphs. These diverse activities required significant amounts of constant funding.

In his quest to maintain private ownership and at least a small degree of private funding of HAO, he appeared to be more influenced by HCO director Harlow Shapley rather than his mentor Menzel. Menzel secured diverse types of governmental and private funding of astronomical and solar research in the immediate post-WWII era. His close association with the Central Radio Propagation Lab and Marcus O'Day of the Air Force makes this clear. However, it appears that by the late 1940s, Menzel had tired of fundraising. He understood that government funding offered immediate relief. Menzel was more than happy then to build Sacramento Peak for the Air Force as long as he could use it. Roberts recalled that Menzel "hadn't had much luck raising money from private sources" to augment HAO's limited budget. As far as finding private sponsors for their ambitious plans for the HAO, "Menzel didn't think it could be done."[6]

Shapley, on the other hand, pushed strongly for the then-proposed National Science Foundation (NSF) as a government sponsored entity to support research at privately owned facilities such as the HCO. As such, he supported the more populist Kilgore version of the proposal as enthusiastically as he opposed to the more elitist Vannevar Bush version of an NSF.

The practical Roberts did not avoid military funding, nor did he wish to at this time; he sought it as vigorously as Menzel did. Roberts worried, however, about "getting under Army control." "Red tape" associated with military contracts particularly bothered him, as he told Shapley.[7] One of the concerns that scientists of this period had was losing control of their scientific disciplines.[8] Roberts never expressed such a sentiment as far as the evidence indicates. He appears rather to have had more mundane reservations about increasing the military and government funding of his work. "I just didn't want to be a government lab and civil servant and all that," he stated, adding that his aversion was most likely just "a prejudice."[9] Without complaint from Roberts, most of HAO's operating and research funds came from the government, and most of that came from military sponsorship. The key issues for Roberts were ownership of the facility and the direction of research.

To get an idea of HAO's funding sources in the late 1940s through the mid-1950s, consider the funding profile in 1954. In this last year of the Harvard–University of Colorado corporation's existence, HAO's operational income totaled approximately $128,000. Of this, $93,000 came from the U.S. government. The Office of Naval Research and the Air Force awarded the bulk of that sum for the coronagraph work at Climax and Sac Peak. NBS (via the Central Radio Lab) contributed $20,000 for continuing the Climax solar data collection work they required for their ionosphere and radio propagation forecasting. Only $36,600, or less than 30 percent of the total, came from private sources.[10] The private sources included the philanthropic Research Corporation, Trans World Airlines (TWA), the Radio Corporation of America (RCA), and other organizations.

The situation was very different when it came to the construction of new facilities, for the government did not directly fund such activities unless they were government assets. Roberts managed to raise an additional $160,000 from private donors for these endeavors. These funding sources provided for the much-needed new observatory structure at Climax and other HAO buildings between 1952 and 1954.

Although this private funding was smaller than government support, it was a crucial component of Roberts's overall funding scheme. The need for Roberts to justify this important, but supplemental, private funding for HAO had more effects on his research and career than did his need to justify government funding. The latter required much less effort to obtain, and he had to develop methods to generate more interest on the part of potential private donors, who represented a much harder sell, as Menzel had come to learn.

An obvious and direct association with broad military interests in sun–earth connection research developed with Roberts's solar work in the war years. Menzel and Roberts helped to establish those links in the first place. The courting of nongovernmental sponsors presented first Menzel and then Roberts with an altogether different and more challenging fundraising context. The private sponsor pool consisted of everything from purely philanthropic institutions to diverse business entities. Roberts therefore required a much different approach in presenting his case for funding to private donors. These potential donors did not have the obvious need to understand the sun–earth environment that the military did. Generic appeals to potential sponsors to help advance scientific frontiers or to better understand the esoteric "sun–earth connection" could not move sponsors effectively; an appeal related to addressing an immediate and widespread concern might. Roberts found that concern in one narrow aspect of sun–earth connection science—the possibility of a sun–weather connection that relates activity on the sun to the age-old and very mundane question of predicting the weather.

The Sun–Weather Connection as a Fundraising Tool

"The eleven year period is not one to be neglected," wrote astrophysicist Sir Norman Lockyer in a 1900 astronomical review referring to the possible linkage between the cyclical changes in sunspot numbers and effects on the earth.[11] The idea that sunspots, and other changes in solar characteristics such as brightness, could affect the earth's weather has a long history. Sir William Herschel, discoverer of the planet Uranus and infrared radiation, speculated in the early 1800s on solar variations affecting earth's climate over centuries. Addressing shorter time scales, he even proposed these variations might affect such things as wheat production.[12]

The discovery in the mid-nineteenth century by Heinrich Schwabe of an approximately ten-year sunspot cycle began a new round of speculation about the sun–earth connection in general, and sun–weather links in particular.[13] Famed naturalist Alexander von Humboldt popularized Schwabe's observations, and speculations on the nature of a possible the sun–weather connection grew. "Sunspottery" some called the subject at the time, and Lockyer soon established himself as Victorian England's most notable and vigorous supporter of the idea that sunspots affected earth's weather.[14]

The controversial notion crossed the Atlantic and took hold in American scientific circles, most notably at the Smithsonian Institution. First Smithsonian director Samuel P. Langley, and then his protégé and successor, Charles

Greeley Abbot, made the quest for a sun–weather connection part of the American scientific scene in the early twentieth century. Abbot cast his stamp on this investigation because of his very long tenures as director of two of America's primary scientific research centers, the Smithsonian Astrophysical Observatory (1906–44) and the Smithsonian Institution (1928–44). As such, his influence helped to shape the research agenda of not only the Smithsonian, but other institutions as well. He focused his search on changes in the brightness of the sun. Astronomers, erroneously Abbot thought, called this parameter the "solar constant." Abbot firmly held to the idea that this "constant" actually varied in time. Therefore, he argued, this changing solar energy output dramatically affected earth's weather. He never relinquished his belief that correlations between changes in the solar constant and earth's weather (and climate) were not only real, but had important consequences for humanity.[15]

Abbot's theories did not find widespread acceptance. They did, however, have an air of plausibility about them, and so the research continued. A number of conferences sponsored by the Carnegie Institution in Washington, D.C., during the 1920s and 1930s encouraged these types of investigations. Victorian era "sunspottery" made something of a comeback in the United States in large part due to Abbot's efforts.

Menzel, and later Roberts, also thought a physical connection between solar variations and earth weather might exist. They also realized that tying their esoteric solar studies to earth's weather gave this research a practical side they could use in courting sponsors. Perhaps the first use of this tactic that combines the quest for astrophysical knowledge with a potential day-to-day application was Menzel's approach to Secretary of Agriculture Henry A. Wallace in 1935 for coronagraph development funds.[16] The success of this approach set the stage for Roberts's fundraising in the 1950s.

Referring to the work of Abbot as a clear "orientation towards the practical uses of science," Roberts "sort of concluded that maybe the most important thing that could come out of solar physics research would be to understand and maybe predict the effects of solar activity on weather."[17] Even as a graduate student, Roberts expressed interest in Abbots's work. Although the trip west may have stimulated his sun–earth connection interest, he developed these interests prior to beginning his journey. He had coauthored a paper with colleagues from Harvard in 1940 on statistical analysis of solar luminosity showing possible variations in the solar constant.[18] Although this line of research took a backseat to doctoral research and his war-related solar work at Climax, the question of sun–weather correlations resurfaced for Roberts at war's end.

This sun–weather focus extended the potential private donor base far beyond the standard philanthropic entities. Using it, Roberts now had entrée to businesses and organizations not usually considered as donors to cutting-edge solar physics. The diverse and disparate list of donors that Roberts could plausibly approach included airline and aviation companies, agricultural concerns, distilleries, and producers of heating and camping equipment.

Roberts employed this approach in the early 1950s when he solicited $250,000 for the buildings at the new, enlarged observatory location about four miles from the original Climax site.[19] The Climax Molybdenum Company planned to expand their mining operations, and this required the demolition of the first observatory. The mining company offered HAO a new site elsewhere on their land. This fit well with the timing of the new coronagraphs that Roberts planned to build with ONR and Air Force money.

In a short, descriptive paper produced for potential HAO donors in January 1952, he wrote that the goal of HAO scientists "is to study not only the fluctuations" of the sun, "but also the related temperature and pressure changes and variations of electrical conditions of the upper layer of the earth's atmosphere." "From researches of this sort," referring to the work going on at HAO, "will come information of vast importance . . . to climatology and weather forecasting if present indications are borne out by later developments."[20]

Roberts wanted to get the message out to potential sponsors that HAO work, although centered on solar observation and study, had a much broader research agenda. Because the sun affected everything on earth, Roberts argued, even those with little or no interest in science for its own sake could benefit from the practical applications of knowledge gained from the HAO. He approached, for example, the Great Northern Paper Company with perhaps some exaggeration in claiming that "much of the basic data that we have accumulated has been a part of the body of information used by meteorologists in the tremendous theoretical advances in the meteorological field in the last nine months."[21]

Sun–weather connection or not, fundraising took up much of Roberts time in the early 1950s, even to the detriment of his scientific work. The strategy of coupling the sun to earth weather worked only to a point. "We had so much trouble raising money," Roberts said of this period.[22]

Roberts recalled that he "pounded the pavements in New York," with Joseph Barker of the Research Corporation helping him find possible donors. Roberts spent, by his accounting, almost a total of six months over the three-year period of 1952–54 in New York City soliciting funds from "hundreds of sources." He offered to these potential donors, as exemplified by his appeal to the Great Northern Paper Company, that it was in their best interest to support his operation since HAO did "fundamental research related to radio communications and to weather."[23] As more and more scientists looked to the government for funding in this postwar era via such entities as ONR and the new National Science Foundation, Roberts remained dedicated to the idea that at least some of his funding should come from private sources, as evidenced by this scurrying for donors. He did this despite this period representing a "bull market" for physicists, in particular with the increase in government funding in the wake of the Korean War.[24]

In this, Roberts held (at least partially) a belief that many astronomers held—private and state sources funded good astronomy.[25] In retrospect, this period of mixed private and public funding represents a transition phase for Roberts; by late in the decade, he would almost totally drop the search for private funding as the U.S. government funded all of his research. Early in the decade, however, the sun–weather approach was successful enough to enable Roberts to enlist sponsors for the new buildings and some other solar research activities from a wide spectrum of American businesses. These included businesses one would not have assumed to have even a remote interest in solar studies going on at HAO. A partial list of contributors to the new observatory includes the Denver Union Stockyard Company, the Colorado Fuel and Iron Corporation, the American Crystal Sugar Company, the Great Western Sugar Company, International Business Machines, Trans World Airlines, and Consolidated Vultee Aircraft.[26]

Roberts not only used the sun–weather connection to help broaden his potential sponsor base, but he used the idea to enlist all the help he could. Frank A. Kemp, head of the Great Western Sugar Company, offered to organize a lunch where Roberts could make a plea for funding at Denver's Brown Palace Hotel in May 1953. Roberts enlisted the aid of University of Colorado president Stearns once again. The university president signed the invitation that told the invitees from the top echelons of the Denver business community that "there is genuine hope that the studies during the next few years will enable the forecasting of long-range weather."[27] Kemp in particular, an HAO administrator wrote to Stearns, expressed interest "solely on the basis of weather prediction possibilities."[28] As early as 1949, Kemp had expressed an interest in HAO's weather-related work to Stearns after receiving a letter from Menzel. Kemp wrote Stearns that "we are peculiarly interested in the development of knowledge or techniques that would support long-range weather forecasts," but understood that this field was "quite obscure" with "a good deal of uncertainty about the outcome."[29] Remaining convinced of the possible benefits of HAO's work in this regard, Kemp enthusiastically supported the drive to gather funds for the new observatory building.

Another enthusiastic HAO supporter from the business world was James L. Breese, president of Breese Burners in Santa Fe, New Mexico. Not only did he attempt to assist Roberts in soliciting funds in 1952 from a major producer of stoves and other camping equipment, the Coleman Lamp and Stove Company, he even outdid Roberts in his appeal. Writing to the Coleman company, Breese stated that "soon television may very well carry a daily digest of yesterday's sunspots . . . together with comments from well-informed scientists with long-term predictions on future weather."[30] The company's establishment, he suggested, of the "Coleman Solar Research Foundation" at Climax therefore could represent a fitting monument for W. C. Coleman, the company's founder. It would cost only, Breese added, "a tenth of your yearly advertising budget."[31] Upon seeing a draft of Breese's three-page letter, Robert's wrote to its author that the missive seemed "perfect!"[32]

The Coleman Company demurred, and Breese "felt terribly" that he had not helped raise a large amount toward the observatory building.[33] He later wrote to the HAO that, in addition to a personal contribution of $2,900, he was "working on a further scheme for future finances."[34] The "scheme" involved Breese giving his entire business to the HAO, perhaps forming a subsidiary corporation. For legal and tax reasons, however, the HAO leadership rejected the surprising offer. Not wanting to pass up on Breese's obvious generosity, one senior HAO administrator did pose an alternative. Perhaps wistfully, he suggested the possibility of getting the businessman to fund "the Breese Coronagraph."[35]

The combination of the diverse private funding sources, foundations, businesses, and individuals resulted in the new observatory at Climax. HAO received support from "37 non-government donors" for the new facility. Roberts described eleven of these as "corporations" and "at least nine contributions were made by individuals who were primarily businessmen" in a letter in response to R. W. Lippman. Lippman in a previous article had presented, according to Roberts, "a rather false notion of the way many business corporations look on the role of basic science in today's world."[36]

Why would businessmen whose interests seemed far removed from sun–earth science want to fund Roberts's work anyway? Far from being provincial and of limited insight into the working of modern science, businesses, Roberts argued, often "make very liberal interpretations of the interests of the companies in judging what is proper for them to support."[37] By tapping into their liberality, he bridged the seemingly unbridgeable divide between the esoteric and the practical, a page from Menzel's book.

Roberts had other reasons to thank businessmen for their "liberal interpretations." In addition to support for basic scientific research of the sun–weather connection such as HAO's, these contributors from the presumably conservative business class overlooked an important incident in Roberts's personal life. At about the same time as his extensive fundraising efforts began, Roberts experienced a U.S. government security clearance investigation in 1950. Discomfiting as it was for Roberts and his family at the time, the security clearance turmoil did not lead to any apparent problems in fundraising—he collected $250,000 in private donations for HAO following the unfortunate episode. The stresses produced by this trying incident resulted in a further increase in the tension between Roberts and his mentor, Donald Menzel.

Roberts Endures the Security Board

Despite the government largess for science at the start of the Korean War, this period was an "Age of Anxiety."[38] Many American scientists, for diverse reasons that often included liberal and leftist views on the world situation, became victims of the post-WWII anti-communist fervor that developed in the nation. Although J. Robert Oppenheimer's 1954 hearing was the most obvious case, many American scientists came under close scrutiny. This was especially true for physicists because of perceived and actual relationships between their research and nuclear weapons.[39] Even scientists with little or no connection to weapons (such as Roberts) came under the suspicion of various government agencies as potential security risks. So this "springtime of Big Physics" was a double-edged sword, for it also brought challenges as scientists assumed a much larger and more visible role in American society.[40]

Few prominent physicists escaped attention. Harlow Shapley and Edward Condon of the NBS both experienced significant anti-communist related travails in the late 1940s. In 1948, the House Un-American Activities Committee (HUAC) labeled NBS director Condon as "one of the weakest links" in U.S. nuclear security.[41] Shapley, not depending on clearances of any sort and having tenure, treated the anti-communist efforts with disdain. He refered to the HUAC as the "Un-American Committee" and often bragged about his defiance of it.[42] The HUAC probably never really wished to target Shapley intensely in any event because he was an astronomer and not an atomic scientist with a security clearance.[43]

Menzel and Roberts's own difficulties followed soon after Condon's. As both were much more junior than Shapley, neither was in a position to treat their situation in the dismissive manner with which Shapley treated his. In addition, Menzel sought increased military support, making him even more vulnerable during this period of increased scrutiny. In Roberts's case, it is important to note how expeditiously he dealt with this difficult issue and how little the experience affected his subsequent work, reputation as an administrator, or fundraising abilities.

This is not to say Roberts assuaged all concerns about his loyalty in 1950, or that all shared a high opinion of Roberts's abilities as a scientist and administrator. Noted director of the Mount Wilson Observatory, Ira Bowen, expressed reservations about Roberts later in the decade. Bowen, however, stated that his concerns in raising these reservations with the Smithsonian's Leonard Carmichael were more about Roberts's "outside activities" detracting from his research than any other question.[44] Irrespective of any issues Bowen had about Roberts's abilities and personal politics, there is no evidence that Roberts's security hearings affected his subsequent ability to raise money for his research or even to obtain security clearances for future government work.[45] There is no evidence that Bowen's reservations about Roberts were widely shared—if they were, it is hard to see how Roberts could have achieved what he did in the decade following the security investigation. Given Roberts's growing reputation, how did he wind up in a hearing in the first place?

Many scientists at the time did research that required clearances.[46] Yet, in an ironic twist from the usual situation where a clearance allows one to engage in classified activities, Roberts did not need a security clearance for his work. He needed it simply to get to work.[47]

Roberts's duties at Sacramento Peak in the late 1940s and early 1950s required him to take a military plane from Lowry Air Force Base in Denver to Holloman Air Force Base at Alamogordo. From Holloman to the observatory, he then had to use a car to transit a part of Holloman where sensitive military activities occurred. One day in early 1950, after he flew in from Denver, Air Force authorities informed him he could not cross the base because his security clearance no longer existed. This made it impossible to continue his work at Sacramento Peak.

Greatly puzzled and disturbed by this episode, Roberts did not get official notification about this problem until September 21, 1950.[48] A review of Roberts's FBI file does not indicate exactly how he came under the scrutiny that led to the denial of the clearance, for names of informants are heavily redacted.[49] It appears that his involvement in activities such as the Cultural and Scientific Conference on World Peace generated suspicions.[50] In any event, his wartime clearance expired, and the government denied a new one.[51] The "nasty stuff," as Roberts called the experience, began.[52]

It is important to note that Roberts did not have to contest his clearance denial. He could have ignored it, with the cessation of his Air Force work at Sacramento Peak as the only immediate consequence to him. There is no evidence that his ONR- and Air Force–sponsored work at HAO would have suffered.

He wrote to Menzel in April 1950 about the clearance problem. Roberts told Menzel that if the situation "increases my difficulties in carrying out an adequate performance of my duties to our projects," then perhaps it was time to "gracefully get somebody else to start taking over my work and for me to start looking for some other job where clearance annoyances will not affect my usefulness."[53] If he had done this and moved on from HAO, not only his career but much of Boulder's development as a center for space and atmospheric science would have altered from what did transpire. He did, however, choose to stay with HAO and fight for what was in reality only a minimally needed clearance.

Perhaps his decision to fight for the clearance centered on two facts. First, it did not appear that Harvard or the University of Colorado had a position for him. He asked President Stearns in July 1950 if there might be a "staff opening . . . where I could retain close touch with the solar program."[54] Apparently, the University of Colorado could not accommodate his request, although the reason remains uncertain.

HCO director Shapley wrote him somewhat obliquely in September 1950 that "both Whipple and I think that a return to the Harvard Observatory would not be satisfactory, and probably would not be possible in view of the University's attitude with respect to terms of service."[55] The final option of staying at HAO and not participating in building Sacramento Peak had no appeal to Roberts.

Shapley, in his letter on the issue, encouraged Roberts to stick it out. "You ask for my own views," he wrote. "I continue to insist, and Whipple seems to agree with me, that it would be a grave mistake for the Harvard Observatory and for the solar work, if you were detached from our western operations." Clearly disregarding Menzel's call by this time for Roberts's termination, the elder astronomer added, "Moreover I feel it would be exceedingly unfair if you were not enabled to continue with the solar project after you have built it up so effectively." He added that "Whipple believes that you should go through the clearance travail. I am not so sure." Shapley gently concluded with "let's get some sleep, some research done, and contemplate the mountains," and closed with a reassuring "you can count on me standing in opposition to any irrationality."[56] Roberts decided to stay at HAO and fight for his clearance.

The formal clearance appeal process for Roberts, although certainly unpleasant, somewhat costly, and stressful, was much shorter in duration than anyone anticipated. In part, the relative ease resulted from Roberts's attitude and approach toward the investigation. A realistic, candid, and methodical approach to the proceedings saved Roberts much grief. As he wrote to Alan Shapley soon after the hearing concluded, "One of the hardest parts, almost, was for me to discipline myself to the realization that I was not going before the board as a man with certain constitutional rights . . . but that I was a person seeking a dispensation." Roberts added that the board was under no obligation to give such a dispensation. A scientist seeking this action, he wrote, went against a review board's tendency to not understand why one "would want to talk to a Russian or other behind-the-iron curtain scientist without being interested in exchanging secrets, or without being stupidly naïve about the real world."[57]

Roberts's hearing before the Industrial Employment Review Board of the Munitions Board began at 10:02 a.m. on December 4, 1950. It ended only a day and half later at 11:35 a.m. on December 5. Roberts, as he told Alan Shapley, was "enormously over-prepared." He obtained numerous affidavits attesting to his loyalty and credibility, wrote an extensive biography, and created a document explaining his actual relationship to questionable organizations with which some accused him of associating. He even compiled a list of statements from noted Americans, such as President Truman, General (later president) Eisenhower, and Air Force chief "Hap" Arnold, comparing their words with comments Roberts made in his own speeches and other public gatherings.[58] He wished to demonstrate his publicly stated opinions on various political issues were not extremist in any way. Journalist and social commentator Norman Cousins served as an expert witness on communism and communist front organizations on behalf of Roberts.

Roberts, who was politically active and had expressed interest in some form of a world government to lessen the possibility of nuclear conflict, successfully showed the board that his thoughts in this regard did not stray at all from the political mainstream. As the transcript of the hearing indicates, he was able also to show he had never been a communist, nor had ever developed any association with the communist front groups that informants had claimed.[59] Roberts told an interviewer in 1983 that he had discovered in the process of preparing for another, higher clearance later in his career that his ordeal transpired in large part because of mistaken identity—the real communist, Roberts claimed, was one "Walter R. Roberts" of Boulder.[60] Government investigators and informants, he surmised, had confused the two.

Irrespective of the real cause of the hearing, Roberts's approach, combined with the services of an experienced Washington, D.C., lawyer, worked. The board quickly exonerated him, opening the way for the approval of his clearance. This positive result for Roberts would have an enduring legacy. It ensured Roberts remained in the city, while also fueling the already disintegrating relationship between Menzel and Roberts. Unbeknownst to Roberts, he had contributed to his mentor's own rapidly developing security problems. The downward spiral of relationships between Harvard and the Colorado group would eventually lead to the disintegration of the HAO Harvard–University of Colorado cooperative effort. Oddly enough, this deteriorating condition would have positive effects on the state of atmospheric and space science in 1950s Boulder.

"I Can No Longer Continue to Recommend Walter Roberts"

"Little did I know," Roberts said reflecting on this difficult time, "that Menzel was in a security hearing."[61] Possibly because of the same Colorado-based informants that accused Roberts of subversive activities, Menzel also came under government scrutiny at almost the same time. Unlike Roberts, who appealed to many of his associates and acquaintances for support in this process, Menzel apparently chose to remain secretive. He informed few about his situation. Roberts had no way of knowing at the time that his testimony directly contradicted Menzel's on a small but important point for the investigators—who invited Soviet scientists to Climax in 1946?

Roberts testified that Menzel was the one who invited the Soviets to Climax, and apparently had documentation to prove it.[62] Menzel stated just the opposite during his hearings, thereby damaging his cause. This perhaps resulted in investigators conducting a more intense scrutiny of Menzel than would have otherwise occurred.

The differing approaches and styles of the men certainly showed up in the manner in which the two cases evolved. Whereas Roberts's troubles lasted less than three months from formal notification to resolution, Menzel's lasted almost three years. This no doubt took a great personal toll on the scientist.[63] The fundamentally apolitical Menzel probably resented his outspoken junior's speedy vindication, especially as Roberts inadvertently contributed to his superior's difficulties. In a draft but apparently unsent letter to Roberts written in July 1950, Menzel states, "You may as well know that my alleged association with you was one of the most serious problems I had to meet."[64] Menzel could not know that Roberts later in the year vehemently denied any suggestion, possibly to his own detriment, during his hearing that his mentor was in any way a "RED" (caps in original).[65]

As early as the spring and summer of 1950, Menzel definitely envisioned a future HAO without Walter Roberts. Writing to University of Colorado president Stearns, he claimed that Roberts's nonscientific pursuits, although "generally encouraged" by Harlow Shapley, had generated "ill-will" against HAO. Further, Roberts "refused to enter actively into the fundraising" for the observatory while his activities, in Menzel's view, discouraged potential donors. Menzel thought, ironically as it turned out, that "I doubt very much that Water Roberts can ever be cleared or . . . that a very long period of time will be required in order to get a full clearance." As a result, he wrote, "I can no longer continue to recommend Walter Roberts for an appointment" as superintendent of the observatory. He planned to inform HAO's trustees of his view as well.[66]

This phase of Roberts and Menzel's increasingly complex and strained relationship began in April, about the time Roberts's clearance issue surfaced. Menzel wrote that some had accused Roberts of "speaking strongly against government clearance" during a talk he had given at a luncheon in Denver, a charge Roberts in turn denied. As an academic and an American, Roberts thought, he had the right to speak on issues of importance to him "even if I am sometimes wrong or misunderstood." He ended his letter by writing that "if doing this is detrimental to my work, then I'm in the wrong job."[67]

After Menzel and Roberts met at Sacramento Peak in June 1950, Roberts wrote Harlow Shapley that "Don suggested that it would be to the general good of the solar plans and programs of the Observatory if I would resign." Menzel, reported Roberts, "would like it all to be friendly and without hard feelings." Irrespective of any feelings Roberts had toward Menzel, this turn of events clearly left Roberts shaken. He thanked the HCO director for being a receptive ear so Roberts could "get it off my chest."[68]

Menzel, according to Roberts, again insisted that his desire to remove him as HAO superintendent resulted not only from Roberts's speaking on international affairs and other controversial political issues, but also from insufficient attention to his observatory responsibilities, including fundraising.[69] Menzel himself seemed of two minds on the question of Roberts's administrative oversight of the HAO—in his July 6 letter to Stearns he admitted that Roberts kept observatory staff morale high and also that he "has done a remarkably fine job from the start and has carried it through over a long period of time."[70]

The HAO trustees were "outraged" when they found out about the exchange between the two men. Menzel's actions particularly perturbed Harvard-appointed trustee Judge William Jackson, a member of the Colorado Supreme Court. Roberts remembered the jurist saying, "'Menzel can't fire you, only the trustees can fire you."[71] In addition, strong support came from Harvard. Shapley wrote Roberts in late June 1950 after word of this controversy between his two subordinates reached him in Cambridge. He advised Roberts to not make any "hasty decisions," as "Menzel is making a very grave mistake for himself, as well as for solar astronomy."[72]

Roberts took the elder astronomer's advice and chose to remain at the helm of HAO in addition to fighting successfully for his security clearance. Menzel appears to not have pushed the issue beyond expressing his desires to Roberts and others such as Stearns. This first crisis passed, in part due to Roberts's speedy exoneration by the clearance review board. It passed also because of the strong backing given to Roberts by all of the leadership of HAO save for Menzel himself. The HAO trustees, President Stearns, and Harvard College Observatory director Harlow Shapley stood squarely behind Roberts. He remained as HAO superintendent, but clearly the relationship between the two astronomers continued on a downward slide despite Menzel's desire to keep things "friendly." The personal tensions led to subsequent bureaucratic tensions between Harvard and its "western station" as well. As Roberts recounted, "thereupon began the separation of Harvard from the High Altitude Observatory."[73]

"My, These People Hate Each Other"

Menzel's attribution of the origins of his clearance problems to his association with Roberts, including Roberts's inadvertent "fingering" of Menzel as the one who invited the soviet astronomers to Climax, did not help this deteriorating relationship in 1950. Another controversial issue relating to observatory governance and management arose in early 1951 and put HAO further on the road away from its Harvard roots and toward independence. This issue came down to one question: who directs the HAO's scientific program?

In late 1950, Harvard provost Paul H. Buck wanted to change the administrative structure of the HAO by recasting its Committee on Scientific

Operations (CSO). The CSO existed from the inception of HAO. Its structure recognized Shapley, Menzel, and Roberts as the guiding troika of the observatory's activities. However, Buck wanted to bring in more scientifically trained voices with interest in solar work into determining HAO's activities and policies. HAO's trustees agreed to the proposal. In addition to Shapley and Menzel as ex-officio representatives, the new CSO members included noted scientists Fred Whipple, Leo Goldberg, Edward Condon, and James Broxon (a cosmic ray physicist at the University of Colorado).[74] Menzel also served as ex-officio chairman of the CSO. Roberts, as HAO superintendent, did not have a seat in the reconstituted CSO.

The new CSO began its existence at a time of growth for HAO. The new coronagraphs, assisting Sacramento Peak, the relocation of the observatory site, ambitious research plans, and the need for new funding required close oversight and administration. The staff alone had grown to almost forty. Because of its Harvard origins, however, the CSO became a Cambridge-centered entity. The CSO therefore assumed an ambiguous role in HAO's administrative structure from its inception—because of its distance from the loci of HAO activities, the Colorado-based HAO trustees, and HAO's superintendent, Walter Roberts. This awkward arrangement proved itself unsatisfactory for all those involved. Menzel and Roberts found in this restructuring yet another cause for friction between them.

Questions arose regarding this new "chain of command." As early as April 1951, Roberts wrote to Menzel expressing concern that the relationships among the trustees, CSO, and superintendent were not clear. He expressed a sense of frustration about this ambiguity, chiding his former mentor with the thought that "you remember that I have been anxious to do this for a long time" in referring to the need for the clarification of these relationships.[75] In response, Menzel voiced his frustration that Roberts "had not been making adequate use of the CSO."[76] This issue continued to eat away at the ties between Harvard and the University of Colorado. It reached a crisis point in 1953 over the proper role of the CSO in the management of HAO scientific personnel. This crisis began the final plunge toward dissolution of the Harvard–Colorado partnership that occurred in the late summer of 1954.

The controversy in early 1953 over the potential hiring of astronomer Isadore Epstein was a symptom of the serious problems existing between Menzel and Roberts as the tensions between them increased further. The dispute started when CSO member Leo Goldberg expressed reservations about Robert's desire to hire Epstein, a young astronomer and student of Martin Schwartzchild. Goldberg used this opportunity also to express his dismay that the scientific staff of HAO suffered from a lack of "distinction." Although Roberts did not mind getting Goldberg's opinions, he did object to the manner in which the CSO, meaning Menzel, handled the situation.[77]

"As I visualize things," Roberts wrote Menzel, "I do not, for example, 'seek permission' of the CSO to deal with Epstein." Roberts thought that the awkward nature of the CSO arrangement often required him to deal directly with his trustees and take other unilateral actions. He did not mind getting "approval" for a job candidate or other forms of CSO advice, but he also "visualized that CSO comment would come informally . . . in the nature of friendly, sympathetic advice—not as evidence of a struggle over control." He ended his letter with the admonition that "it is I, not the CSO, on whom will fall the blame if the decision is wrong."[78]

"You have relegated us to an advisory role," Menzel responded, although he conceded Roberts's point that the "dispersion of the CSO preclude its operating with highest efficiency." He stressed his view that "CSO should have control in matters of scientific policy and . . . major staff appointments." Indicating some of the confusion Roberts thought existed in the CSO's role, Menzel added, "Of course, the CSO cannot force you to hire someone you do not want on your staff or to undertake some research program." The CSO only wanted, in Menzel's understanding, "to help take some of the responsibility in the making of decisions." As Menzel viewed his actions as benign, he told Roberts that "I just do not understand your point of view."[79]

This lack of communication between the two scientists caused the relationship to further spiral downward. "I was not consciously aware of the fact that your letters were intended to wipe me off the map," Menzel wrote Roberts in April. As Roberts wanted to "make all of the decisions," the CSO's role became irrelevant in Menzel's view. Therefore, Menzel would "advise discontinuing it entirely."[80]

"My, these people hate each other," commented University of Colorado sociologist Howard Higman to HAO business manager Robert J. Low upon reviewing, apparently on Low's request, the Roberts–Menzel correspondence relating to the CSO controversy. Low and Higman reached insightful conclusions about the deteriorating relationship between Menzel and Roberts.

"WOR . . . is trying to destroy CSO," but Roberts also "has a father complex about Harvard, CSO, and Menzel." Menzel, for his part, "is suffering from a guilt complex about his association with CSO" and is chairman of CSO "for no other reason that that he used to be director of the whole outfit." Menzel became acting director of HCO in October 1952 and had to further remove himself from HAO operations and oversight. The "guilt complex" resulted from him having a key place in the HAO hierarchy, but doing nothing "for HAO in the past two or three years that would warrant him having such a position."[81]

"The Harvard part of the relationship between the two organizations is becoming more and more one of performing a historical ceremony. CSO is an anachronism," they concluded. The question became for them, and implicitly Roberts, "how we dump the CSO."[82] The opportunity to "dump" both CSO and the Harvard connection came in early 1954. Despite Roberts's assertions years later that he did not want the breakup, he for the most part led the charge.[83]

More and more of those involved, by later 1953 and early 1954, questioned the point of Harvard's continuing involvement in HAO. Many familiar with HAO, in Colorado especially, gradually came to think that Harvard (via Menzel as the CSO chairman) wanted to maintain control of HAO while having very little vested in the observatory's activities other than its name. The cumbersome relationship among the CSO, HAO trustees, and Roberts became increasingly unworkable, with the lingering effects of Menzel and Roberts's security clearances travails no doubt complicating the situation. All this led to a crucial meeting on the future of HAO at Sacramento Peak on February 17, 1954, among Menzel, Roberts, and John Evans. Evans, perhaps serving on this day as a referee, moved from HAO to become director of Sacramento Peak Observatory after the Air Force assumed day-to-day operations of the observatory in 1952. As a result of these negotiations about HAO and the CSO Roberts reported to the new University of Colorado president Ward Darley.[84]

After outlining the discussions and his views regarding the interests of Harvard, the University of Colorado, and HAO, Roberts wrote that he and Menzel agreed that if Harvard–Colorado ties remained, the CSO's role "should be advisory to the trustees, but with some responsibility delegated to CSO." They also agreed "that it is probably preferable, if CU is interested, to undertake a friendly separation of Harvard from the enterprise," with all responsibility for HAO going to the University of Colorado. Roberts eagerly recommended to the university president that "steps be taken to find a workable agreement for bringing HAO into the University of Colorado."[85] Only a few weeks later, Roberts wrote and forwarded a draft of the "Plan to Bring HAO into the University" to Darley.

Darley acted quickly. He sent a letter to Harvard's dean, McGeorge Bundy, asserting that the University of Colorado stood ready to assume full responsibility for HAO. Bundy wrote that he "followed the discussions between Don Menzel and Walter Roberts," and had "come to feel that this would be the ideal solution."[86] Harvard did not mind being "dumped," as Low had called the possible breakup. From the tenor of Bundy's letter, it appears that the eastern university looked forward to it, and did not wish to think further about HAO.

A number of causes led to Bundy's speedy and enthusiastic acceptance of the Colorado offer. He elucidated these for Darley in a subsequent missive on August 6. This letter actually came after the HAO trustees, the approving body for the corporation, expressed concern about the planned separation from Harvard. The concern centered on the damage to Harvard's reputation in Colorado, not damage to the University of Colorado or HAO. According to one trustee who wrote to Bundy, "the regard with which Harvard is held in many parts of this country is deplorably low."[87] To assuage the HAO board, Darley in July sent Bundy a revised offer that would keep Harvard in the fold, but in effect, on Roberts's and the University of Colorado's terms.

Bundy stated the "embarrassing fact" that Harvard's inadequate financial resources for astronomy required something of a drawdown of its far flung astronomical interests, including their field site at Bloemfontein in South Africa. More to the point, he acknowledged what many people such as Low and Higman had already observed—the situation surrounding HAO had changed much since 1946. Bundy observed that the increasingly autonomous HAO "has grown up and wishes to be scientifically independent." As Harvard's association "has been inextricably connected" to the CSO as a "strong policy-making" group, any suggestion that CSO be solely advisory necessitated the split.[88] Bundy then inadvertently generated some additional controversy, especially with Harvard-appointed HAO trustee Knowles and former university president Stearns (who had moved to the Boettcher Foundation in Denver). The Harvard dean emphasized that his university really wanted the CSO as an oversight body because Roberts "did not seem to me to be an astronomer of the first rank."[89]

Perhaps the fundamental reason for the breakup remained unstated in this August correspondence. A few months earlier, Bundy called Harvard-appointed trustee Thomas Knowles about the looming breakup. As Knowles wrote Darley soon after, "the basic reason" for Harvard's action in this "divorce" is an "irreconcilable conflict between Roberts and Menzel."[90] The problems that had possibly begun with a difference in the best way to fund HAO, exacerbated by "the strain of the security business" as Roberts later claimed, and complicated by Roberts's maturation as a scientific administrator, led to an inevitable conclusion.[91] Many in Boulder saw this separation coming over the years and wanted it by 1954, as the candid Low memo of 1953 indicates.

Bundy finished his letter of "divorce," to use trustee Knowles's description of the separation, with "it is very good news . . . that our friends in Denver and Boulder do not want us to withdraw." Harvard, despite this sentiment in Colorado, "must end its official responsibility for the Observatory." Bundy then came full circle and referred to Darley's original proposal letter of March 22 when asserting that it was in everyone's "best interest" to dissolve the corporation.[92] The eight-year-old corporation for astronomy that no one had ever seen the likes of before ended. This break also ended Roberts's drive for increasing autonomy from Harvard and Menzel. As Roberts cheerily wrote to Hugh Odishaw of the U.S. National Committee for the International Geophysical Year, "smoother functioning" would result from the split and the restructured HAO corporation since "we agree with ourselves more often and more easily than with others!"[93]

In October, Menzel wrote Roberts after receiving the notice of the Harvard corporation's decision to withdraw from HAO. With a dramatic change of tone from previous correspondence, as if relieved the long personal struggle between the two men had finally ended, he noted that "all of us view this action with mixed emotions" and "our joint scientific work over the last several years has reached a new and continuing high." It appears Menzel wished to put the intense personal conflict behind them. He looked forward to interacting with Roberts and HAO in the newly formed group of solar researchers called the "Solar Associates." Indicating the complexity of this lifelong relationship, Menzel concluded with a seemingly heartfelt "best personal regards and every wish for success."[94] Harvard and Menzel no longer had any direct influence in Boulder. HAO, from thereon, was free to determine its own fortunes.

The HAO trustees had wanted to keep the corporation as it was; they eventually accepted the inevitable. In this process, the Harvard-appointed trustees expressed their strong dislike for the way Menzel had handled HAO and attributed the "divorce" ultimately to him. "It is clear to me," wrote Knowles to Bundy, "that this move on Harvard's part has been provoked by Menzel as some phase of a personal contest with Walter Roberts." Knowles stated he had "no interest in being unfair about it," but perhaps in a way he was.[95] As Low's memo of a year earlier demonstrated, it appears that senior HAO staff had pushed for "dumping" the CSO and Harvard much earlier.

Knowles perhaps was inadvertently unfair in another way as well. The separation of HAO from far-away Harvard seems, in retrospect, understandable. Starting with the Climax operation, HAO began in effect as a "colony" of U.S. astronomy's eastern establishment HCO as part of Menzel's and Shapley's "western stations."

Understanding the spread of science (from the Northeast to the Southwest) in this postwar era as a type of scientific colonization is applicable not only to Boulder's story. Some scholars have stated this was a major, although often underappreciated, theme of the development of U.S. science in this period—the Southwest was colonized by the northeastern national scientific establishment.[96] The (Harvard) colony in Boulder thrived and matured under Roberts's work and leadership and by 1954 had established a "colony" of its own with the Air Force–sponsored Sacramento Peak site in New Mexico.[97]

The developmental process of HAO in Boulder fits well as an analog to the process of colonization—a process that often leads to political independence. As described previously, this was precisely the pattern events in Boulder followed. Roberts felt increasingly alienated from far-away Harvard (and Menzel), and as a result wanted more autonomy. This process eventually led to a split that gave much greater freedom of action to Roberts.

As such, HAO in Boulder forms an integral part of the story of U.S. science as it moved from its eastern roots to other regions—the "scientific Siberias" of the country. Time eventually severed those roots, with significant implications for the development, in this case, of Boulder. As a result of the split, fresh opportunities, along with many challenges, confronted the newly independent HAO and Walter Roberts.

Roberts, the Sun–Weather Connection, and "Astro-Geophysics"

As HAO ties with Harvard dissipated, ties grew stronger with the nearby University of Colorado. Similarly, as HAO interests diverged from Harvard, they converged more and more with the ambitious plans a series of University of Colorado presidents developed for science at their school. The plans began with Stearns, accelerated with Darley, and reached a high point in the mid- and late 1950s with President Quigg Newton. The observatory (and Roberts's) roots also deepened in the local community—HAO and the small city of Boulder by the mid-1950s had become inseparable. Roberts later spoke of this time as the period when "the observatory really began to go forward and blossom."[98] The creation of the Institute for Solar–Terrestrial Relations (ISTR) was one of the ways HAO moved forward in this new phase of its existence. ISTR also had the important effects of drawing HAO closer to the university and of getting Roberts more involved with U.S. atmospheric scientists and meteorologists.

Roberts wrote Darley in September of 1954 suggesting that HAO come under the university and its regents as a separate "astrophysical institute" not attached to any one department. The new "High Altitude Observatory of the University of Colorado" could then serve as an academic asset contributing to all departments and educational activities of the school as appropriate.[99] After some discussion, and driven by legal considerations, the original corporation continued with amended bylaws. The university president could appoint all the trustees under this new arrangement, and the regents agreed to Roberts's suggested name for the reorganized observatory. On February 17, 1955, the HAO trustees amended the certificate of incorporation and approved the new name.[100] A new phase for both the HAO and science in Boulder began as Roberts (and HAO) struck out on their own path under the auspices of the University of Colorado. The first major effort for HAO after gaining its autonomy was the ISTR.

"That was really the beginning of major meteorological research here," Roberts said of the ISTR project years later and its significance to Boulder.[101] Roberts's interests in the sun–weather aspect of the broader topic of sun–earth relationships had captivated him since his graduate school days. Wanting to create this new center as part of HAO to broaden and deepen its intellectual roots, Roberts continued with the technique that had worked well in the quest for the new observatory buildings at Climax—fundraising from private sources. Again, he used the tactic of soliciting funds from possible donors based on the potential practical benefits of sun–weather research.

For Roberts, private funding constituted an important part of how science should be done. In addition to avoiding the dreaded paperwork associated with the military contract system, nongovernmental funds also helped to provide for the continuity of HAO by establishing increased levels of financial stability. Piled on to the bureaucratic issues coming with them, the government contract funds from the ONR, Air Force, and NBS barely met day-to-day expenses. Sometimes the contract funds could not even cover these. The termination of these contracts always loomed as a threat to HAO's work as well.

Roberts's ISTR plan called for a fixed time limit on research. He allocated four years to describing and explaining the mechanics by which transient solar activity such as flares might affect daily terrestrial weather. His previous pleas to HAO donors suggested in only a general way how their contributions to HAO solar work could help scientists better understand the solar processes that might affect the earth. The ISTR plea focused on the support of HAO sun–weather research. HAO and Walter Roberts therefore took the plunge into atmospheric science, albeit from an oblique angle.

"We have made good progress in establishing the reality and importance of solar effects on the strength and direction of wind flow in the higher atmosphere," Roberts claimed to important aviation consultant and financier William A. M. Burden in September 1955.[102] Using this approach, in less than a year he garnered almost $150,000 from entities as diverse as the American Gas Association, a number of oil companies, RCA, the Sloan Foundation, and United Aircraft. Roberts's arguments on the possible practical importance of research into the effects of short-term solar variability on weather also swayed individual major philanthropists such as Laurence Rockefeller. Roberts did not stop here. He wrote to a friend that "I am working hard on Jersey Standard, Socony, and Gulf—and I think the signs are favorable."[103] Roberts left few major oil and gas companies untapped in his search for ISTR and observatory funding.

Part of this sun–weather work consisted of collaboration with Richard Craig and Ralph Shapiro of the Air Force Cambridge Research Center. He did not shy from exploiting this association in the quest for donors with strong government ties. Shapiro, he wrote to vice president of Douglas Aircraft Frederick W. Conant in 1956, undertook a study based on "a suggestion from us" that possible sun–weather effects might relate to the surface solar activity that also caused geomagnetic disturbances.[104]

He also wrote to the Douglas executive that he was "happy to say" the proportion of federal funding had declined in the previous few years to only about half of HAO's budget. This fit with his view that some funds for his work should come from nongovernment donors. As "happy" as he was about the decline, he added that it therefore necessitated more support from private sources such as Douglas Aircraft. Such funding, Roberts emphasized, "greatly strengthens . . . the scientific program of the Observatory . . . precisely because it is supported in such large measure by private industry."[105]

As part of the ISTR sun–weather effort, Roberts had to broaden his interactions with like-minded meteorologists and atmospheric scientists. In addition to his work with Shapiro and Craig of the Air Force lab, Roberts increased HAO's meteorological expertise. He convinced noted atmospheric scientist Bernhard Haurwitz and Julius London from the strong meteorology program at New York University to join his operation in Boulder. He also invited Herbert Riehl, a widely recognized founder of atmospheric dynamics from the University of Chicago, and Hurd Willett of MIT to participate in ISTR activities and seminars.

By bringing in these atmospheric scientific heavyweights into the HAO fold, Roberts tried to dispel the perception of many meteorologists that sun–weather work represented a form of "witchcraft."[106] Many meteorologists of the time thought of pursuing this line of inquiry as "undignified."[107] However, enough scientists did think the issue at least worthy of some study. As a result, important scientific and funding organizations such as the American Meteorological Society and the NSF took note of Roberts's efforts. They agreed to cosponsor some of the seminars run by ISTR. This work broadened the exposure HAO, Roberts, and Boulder had to scientists other than those doing just solar astronomy.

Roberts never viewed the entry into atmospheric science via the sun–weather work as a segue into other scientific disciplines, nor did he see himself in the creation of a new discipline. Rather, for him, this work resulted from a simple "accretion" of both his diverse interests and a long-held desire to put science in service to mankind.[108] The fact that Roberts used the sun–weather connection, as did Menzel before him, to generate sponsorship should not lead one to believe cynically that they did not believe such a relationship existed. Both, on the contrary, thought such a linkage between the sun and earth's atmosphere physically possible. In a 1949 text on solar physics, Menzel wrote in all sincerity that "in the future, it seems that solar studies will play a more active part than ever before in the forecasting of weather."[109]

Roberts did understand quite consciously that this "accretion" of interest had put him on a new path of scientific research. "I also visualize," he wrote to Leonard Carmichael of the Smithsonian Institution in early 1955, "a program of solar-terrestrial research of a type that nobody now is carrying out—that could break into a new field of research more or less at its start."[110] Later that year, in a correspondence to U.S. Weather Bureau chief Francis Reichelderfer, he referred to this new field using the awkward and ambiguous term "solar-meteorological research."[111]

Roberts settled on a name to this research that "nobody now is carrying out." This work consisted of his blend of solar research and atmospheric sciences. Possibly taking a page from his former mentor, he started referring to his work about this time as "astro-geophysics." Menzel roughly defined the term as "the interesting zone that lies between geophysics and astrophysics."[112] Although the sun–weather line of research was not new—William Herschel did it at the turn of the eighteenth century, Lockyer in the nineteenth, and Abbot in the twentieth—Roberts wished to place a new and stronger emphasis on this sun–weather aspect of the larger sun–earth connection field. Even before Menzel, HCO director Harlow Shapley had proposed such interdisciplinary research and continued his interests throughout his career as a supplement to his astronomical work at HCO.[113] Boulder sun–earth connection science was not merely a stepchild of astrophysical studies or part of a wider astronomical research program. For Roberts, and many others, sun–earth research was the most important focus of Boulder science in the city during the early and mid-1950s.

Earlier in 1954, Menzel began teaching a course at Harvard called "Astronomical Geophysics," although in this course there was little or no mention of the sun–weather connection.[114] By 1956, it appears that both Menzel and Roberts began using the term "astro-geophysics" to describe their different approaches to what at the time seemed to many in their discipline as "non-traditional astronomy."[115] This was an era of fragmentation of astronomy into subfields, as indicated by Whipple's "New Horizons" panel at Princeton in 1955. The group identified many new areas of astronomy other than galactic and stellar research that might qualify for NSF funding—"solar system science" among them.[116] Roberts's support for such studies therefore served to distance him from not only some of his colleagues in traditional astronomy, but from solar researchers well. As if to balance this increasing distance from the astronomical world, his interdisciplinary interest pushed him closer to the nation's atmospheric scientists and meteorologists.[117]

After ISTR began work in early 1956, Roberts persuaded University of Colorado president Darley and the regents to approve the creation of the Department of Astro-Geophysics to augment and support the new institute. This was about the same time that Menzel started using the term at Harvard for some of his work.[118] Roberts's motivation to create this new department perhaps also related to a major reason the University of Colorado and HAO lost out to Harvard in 1955 as the new home of the relocating Smithsonian Astrophysical Observatory (SAO)—Colorado lacked graduate training in astrophysics.[119] The faculty committee created to study the proposal reported on the obvious reality in Boulder that Roberts thought the time had come for such a department: "The facilities at present in Boulder to a graduate student in astro-geophysics are quite considerable, and prospective expansion makes them even more favorable."[120] Although Boulder was not yet a powerhouse of astrophysical research, it was a significant place for "non-traditional" sun–earth studies by the mid-1950s.

The University of Colorado had a poor standing in academic circles—almost an intellectual backwater—separated as it was from the eastern educational establishment by half a continent. Roberts recalled that the university in the 1950s "didn't have a very good reputation in those days for science . . .

we were out of our class" when speaking of the failure to get SAO moved to the school.[121] He clearly and consciously tried to alter this perception by helping to establish the new department as a close adjunct to HAO. Roberts served as head of this new academic unit and a professor at the University of Colorado, upgrading from his previous title of research associate. By Roberts's actions then, another important site for generating knowledge about space and atmospheric science had arisen in Boulder.

Boulder as a City of Knowledge

By 1956 Boulder had advanced toward becoming a major city of scientific knowledge production. A basic and informal community grew in the city of diverse but interrelated scientific activities, with the better understanding of the myriad physical processes connecting the sun and earth as the nexus of these activities. The HAO Institute for Solar Terrestrial Research, University of Colorado (via the Astro-Geophysics Department, the Upper Air Lab, and the Sommers–Bausch Observatory), and the NBS Central Radio Propagation Lab provided the foundation for this nexus of sun–earth work in the city. The research and graduate students generated from these new institutions put Boulder on the map of not only U.S. but international science as well.

Circumstances had converged to allow Roberts to achieve an interesting professional feat: he managed to raise his stature in the U.S. scientific establishment despite his embrace of nontraditional interdisciplinary studies, in particular the aforementioned sun–weather "witchcraft."

Because of the many interdisciplinary facets of sun–earth science, this process of Boulder's continued development as a city of scientific knowledge production was natural. Sun–earth science at its core represented many disciplines, so the city naturally began to attract practitioners of these other disciplines. Boulder thus became home to many areas of scientific research when the opportunities arose to bring more scientific research (and technology) to the city as the years progressed.

As the Smithsonian's decision not to move the Smithsonian Astrophysical Observatory to Colorado showed, Boulder had not yet fully developed as a place for international science. The impetus for further progress came from the continued and rapid development of the unique array of scientific assets that then constituted the Boulder community for space and atmospheric science research. Circumstances, with no centralized planning of any sort, positioned the city between the mid-1940s and the mid-1950s to become a major front in what one journalist called an "assault on the unknown"—the International Geophysical Year of 1957–58.[122] This assault (and the response to the IGY-inspired *Sputnik* launch in 1957) would boost the small Rocky Mountain foothill city on its trajectory to becoming an internationally recognized center for the production of scientific knowledge.

CHAPTER SIX

Global Science in One Place

The mid-1950s presented a time of challenge and opportunity for Roberts and science in Boulder. With the newfound autonomy generated by the "divorce" from Harvard and Menzel, Roberts could venture out on scientific endeavors without the worry that his overseers back east at the HCO might disapprove. He had the backing of the HAO trustees and the University of Colorado president, Ward Darley.

The broader context for the next set of developments in Boulder began in the early 1950s with the idea that nations of the world should engage in a global, interdisciplinary, and rare scientific investigation of the entire planet earth and its space environment out to the sun. Called the International Geophysical Year (IGY), this effort encompassed more than sixty nations across the globe, thousands of reporting stations, and 60,000 scientists.[1] This coordinated and ambitious program used airplanes, balloons, ships, and notably, rocket and satellite technologies to obtain global data.[2] Boulder entered into the world of big American science by benefiting from what in many ways exemplified modern big science, the IGY. Both the preparations for the IGY and the dramatic shifts in U.S. science policy that resulted from it powered Boulder's ascendancy to a world-class city of scientific knowledge production.

Many discussions of big science focus on the "vertical dimensions" of integrated "masses of people and material in one place" and therefore have produced a perception that this is what big science primarily consists of.[3] The IGY, by contrast, represented a global network of scientific activity connected by diverse means that encompassed not only the earth, but well beyond the planet—big science as "large volume." Sciences dealing with the natural world, such as astronomy, atmospheric physics, or sun–earth connection research, may have big science aspects built into them.

To support its ambitious efforts, the IGY program developed centers of scientific activity, many of which continued beyond the formal IGY period. Some of these centers came to Boulder. This chapter outlines this process and shows the many interlacing levels that made up U.S. science in this era—personal, local, national, and international. Allan Needell pointed out that "established networks, associations, and visions" served an essential role in producing U.S. science policy in this era.[4] The situation with Roberts, Alan Shapley, and others in bringing more science to Boulder clearly bears this observation out.

It is best to avoiding being "dazzled by the sheer bigness of it all," referring to the IGY and its effects on Boulder, as some critics of big science historiography have argued.[5] Instead, we will see how Boulder science grew because of the IGY, becoming an important site for the production of global knowledge.

Origins of the IGY and the Polar Years

The genesis of the IGY may date from nineteenth-century naturalist Baron von Humboldt's ideas about the need for global data in order to study the complex physical linkages that make up the earth and its natural environment. Boulder's rapid transformation as a site for space and atmospheric science via IGY-sponsored research rests, in part, on a foundation consisting of a seven-layer chocolate cake. James Van Allen, one of the creators of the IGY, often gave credit to his wife's baking skills for the "high spirited" discussion in his home on April 5, 1950, that led to the IGY proposal. According to Van Allen, during a "bull session" enlivened by the aforementioned cake and brandy, Lloyd Berkner turned to pioneering magnetospheric physicist Sydney Chapman and asked the noted researcher if he thought "it is about time for another international polar year?" The influential Chapman not only responded in the affirmative, but also added that he had "been thinking along the same lines myself."[6] The notion of a major investigation of the earth–sun system was in many scientists' minds, generated by the desire to use new scientific tools such as rockets to investigate previously inaccessible regions of the earth's atmosphere and outer space.[7]

German naturalist Friedrich Heinrich Alexander Baron von Humboldt stood among the first in the nineteenth century to call for global measurements of natural phenomena, especially of earth's magnetic field.[8] In time, and following the example set by von Humboldt, the first polar year of 1882–83 began as a suggestion in 1875 made by Lieutenant Karl Weyprecht, a polar explorer and Austrian naval officer.

Thinking that solitary polar expeditions alone could not generate true scientific understanding of these regions, Weyprecht decided that a year-long cooperative effort among nations to gather scientific data might be useful. By doing away with competition and encouraging the sharing of results, Weyprecht reasoned, knowledge of polar regions could greatly increase. Approval of his idea came by way of the first International Polar Conference in 1879. Twelve countries, including the United States, eventually participated. Much data resulted from coordinated observations of atmospheric and geophysical phenomenon. These data included extensive observations of the aurora, geomagnetism, solar radiation, and meteorological variables.[9]

The United States participated in the first International Polar Year (IPY) via the U.S. Army. The nation also participated in a smaller way during the second polar year in Depression-era 1932–33. John A. Fleming, director of the Carnegie Institution of Washington's Department of Terrestrial Magnetism (DTM), convinced the U.S. Congress to support this international activity based on the potential practical benefits of such investigations. Among the reasons he cited included the need to better understand the polar geophysical environment for communication purposes.[10] Berkner, leaving a position at the National Bureau of Standards, joined DTM and Fleming in July 1933. He therefore overlapped with DTM's IPY activities and saw them first hand. As a radio propagation specialist and ionospheric physicist, Berkner clearly understood the importance of these polar years in generating not only basic scientific knowledge about the global environment, but also the practical uses of that knowledge.

When he proposed the idea of a new IPY at the Van Allen home in 1950, Berkner may have had other benefits in mind other than simply the pursuit of scientific knowledge and the practical application of this knowledge. In a move inspired by his long-time friend and then Undersecretary of State James E. Webb, Berkner took a leave of absence in 1949 from DTM to serve as a scientific advisor the U.S. Department of State.[11] In this capacity, Berkner headed the effort to create a report on the role of science in U.S. foreign policy. The report emphasized that science could further international understanding and cooperation. The report also indicated the usefulness of scientific endeavors to gain information about scientific activities in other nations as well.[12] He wrote in 1954 that "tired of war and dissention, men of all nations have turned to 'Mother Earth' for a common effort."[13] This statement may reflect simple rhetoric on Berkner's part, given his interests in the national-security related aspects of global science and the opportunities international science presented in advancing American interest in the postwar era.[14] It may also reflect his experience in the increasingly interconnected world of geophysics and sun–earth science, where cooperation was vital. The many international committees, some discussed subsequently, that formed in the nineteenth and twentieth centuries demonstrated the increasing cooperation that modern geophysics required. It appears that Berkner had a complex set of reasons that are inseparable when making the IPY suggestion to Chapman in April 1950.[15]

Irrespective of the range of motives for Berkner's proposal and eventually the U.S. government's participation in a new IPY, the idea of such an international collaborative scientific activity quickly gained wide acceptance in the United States and internationally. The matrix of scientific organizations that formed the basis for the IGY also played an important role in making Boulder a place for IGY science. These organizations provided a ready set of planet-wide contacts that helped to make Boulder a key site for IGY science.

The Mixed Commission on the Ionosphere of the International Scientific Radio Union (often given with the acronym from the French translation, URSI), the International Union of Geodesy and Geophysics (IUGG), and the International Astronomical Union (IAU) meeting in Brussels in 1950 recommended to its sponsoring unions that a third IPY could yield great returns. "The idea caught on quickly," Berkner later wrote. The overarching International Council of Scientific Unions soon appointed a "Comité Spécial de l'Année Géophysique Internationale" (CSAGI) to plan and coordinate the IGY. Sydney Chapman, described by Berkner as the "world's most distinguished geophysicist," served as CSAGI's president.[16] Chapman, an inveterate world traveler and itinerant scientist, also had close ties to Walter Roberts because of a large overlap in their scientific interests—especially in the sun–earth connection.

Berkner, still president of Associated Universities, Inc., found time to assist his good friend and co-instigator of the IGY as CSAGI's vice chair. Each participating nation created national committees to support the IGY effort. The relatively new National Science Foundation, as its first major international project, funded U.S. participation.[17] The National Academy of Sciences (NAS) National Research Council (NRC) directed the U.S. National

Committee (USNC), which ran the national effort.[18] Joseph Kaplan, an atmospheric chemist at UCLA, headed the USNC. At the USNC's first meeting in March 1953, the members unanimously elected Alan Shapley of the NBS in Boulder, then only thirty-four years old, as vice-chair of this high-level U.S. scientific policymaking body.[19] Hugh Odishaw, Edward Condon's assistant at the NBS and a figure in the CRPL move to Boulder, became the full-time USNC executive secretary. He served as the most important administrator of U.S. IGY activities. According to one IGY participant, Odishaw was the "the guiding genius" behind U.S. IGY participation despite a "dark, moody, and sometimes abrasive personality."[20]

The designated start for the IGY was twenty-five years from the International Polar Year 1932. This coincided nicely with the predicted maximum of activity in the approximately eleven-year sunspot cycle. The previous polar year in 1932 on the fiftieth anniversary of the first IPY had no such fortune: the timing was such that this polar year occurred during a solar activity minimum, decreasing the scientific results associated with the sun–earth connection. The IGY, set during a time of maximum solar activity, promised a more interesting scientific yield. This timing not only guaranteed better science, but more notable headlines than the previous polar years had garnered.

Because of a focus on investigating the entire earth and a desire to better investigate the sun–earth connection in all its dimensions, participants then redesigned the IPY as the International Geophysical Year (IGY). The name change, wrote Berkner, "reflects recognition of the need for world-wide synoptic observation and analysis."[21] The name change broadened potential interest in the project, especially in the United States. Some scientists, such as DTM's Merle Tuve, who disapproved of large-scale and overly directed science, expressed skepticism about the IGY due to the big science aspects inherent in the effort.[22]

Because of the ambitious nature of the research plans developed by the participants, and to take full advantage of the forecasted solar maximum period, the "year" became a year and a half, extending from July 1, 1957, to December 31, 1958. Once underway, this ambitious scientific endeavor gave Boulder a major ingress into the rapidly developing world of international sun–earth science.

Bringing the IGY to Boulder

The IGY, by its very nature, emphasized research in a global context. As Weyprecht realized in the nineteenth century, understanding of the earth and the earth's environment in space required global knowledge generated by participants from many diverse locations using a host of both in situ and remote-sensing techniques. Yet, even as the phenomenon or region of study may be global—"the atmosphere" or "outer space," for example—eventually the study and archiving of data obtained globally must occur in discrete locations.

Many IGY researchers viewed Boulder as a potential, if not obvious, candidate to become one of these specialized centers for IGY activities. Both the scientists who had migrated to Boulder and the institutions that developed there in this period made Boulder a desirable choice as a nexus of IGY activities.

Roberts, and then Alan Shapley after he moved to Boulder in 1954, understood the opportunities the IGY afforded not only their science, but also Boulder. Some scientific communities—oceanographers, for example—had diverse opinions about the value of international science in this era.[23] It appears none of Roberts's colleagues in Boulder harbored any such reservations about international research. All in the city understood sun–earth science by its very nature was international in scope. Roberts and Shapley worked closely together, along with local colleagues, over the IGY period. During the IGY's planning, execution, and post-event phases, they exploited many opportunities for the advancement of both their own organizations and the city of Boulder.

Roberts and Shapley, because of their particular scientific interests and positions as administrators and scientific entrepreneurs, played particularly important roles in the IGY almost from its formal beginning in 1953. Scientists with interests in solar–terrestrial physics represented a small community at this time. Roberts and Shapley's research agendas not only coincided with many of the broad goals of the IGY, but also with CSAGI president Sydney Chapman's research interests. Chapman, a widely acknowledged pioneer in the disciplines of magnetospheric and upper atmosphere physics, formed and maintained close working relationships with both HAO and CRPL and routinely traveled to Boulder. All three scientists shared the bond of sun–earth connection research.

In early 1956, Roberts convinced Chapman "to represent HAO . . . at all times" during his frequent travels and to spend four months a year in Boulder.[24] Roberts and Chapman formed such a close and lasting friendship that the American Geophysical Union (AGU), after deciding to give the prestigious William Bowie award for lifetime achievement to Chapman in 1962, chose Roberts to write the citation and present the medal.[25]

Roberts's role as a key player in the IGY fits into the global scientific community that comprised IGY activities. Kaplan, after the start of formal

IGY activities, wrote to Roberts and asked if he would serve on the USNC's Technical Panel for Solar Activity. Roberts was well known for his solar (and sun–earth) interests, and Kaplan said his help was "urgently needed."[26] Roberts responded that he was already familiar with the IGY and its programs and was therefore pleased to accept the offer.[27] Roberts soon assumed the chairmanship of the solar panel. According to the panel's secretary at the time, Stanley Ruttenberg, the group "hardly ever met" because Roberts had set up the solar program so efficiently and well.[28] Through the formal U.S. IGY program, Roberts achieved even more notice as a sun–earth researcher and capable science administrator. Often in the IGY, as Roberts's case shows, these formal associations and arrangements closely followed previously established social and professional ones.

As was the case with Roberts, Alan Shapley's close connection to IGY scientists and administrators demonstrates Needell's observation that previously long-established personal relationships, scientific reputations, and shared visions played crucial roles in the development of postwar science policy.[29] These associations and resulting intellectual communities based on shared interests form an important component of what some researchers refer to as an "ecology of knowledge."[30]

Needell's observation applies to development of Boulder as a center for the production of scientific knowledge. Alan Shapley's close association with Roberts dated from the 1930s. It remained continuous until Roberts's death in 1990. Despite his relative youth, many guiding IGY figures knew the younger Shapley well because of his Harvard background, noted astronomer father Harlow, and important solar–terrestrial physics work at CRPL. It fell upon Alan Shapley, along with Berkner and Chapman, to undertake the key assignment of spreading the word about the proposed new IGY at various international scientific meetings.[31]

Alan Shapley likely knew Hugh Odishaw from the efforts to relocate the NBS labs to Boulder in 1949–50.[32] On a suggestion from the NSF, he prevailed upon Odishaw to serve on the USNC as its chief administrator.[33] Similarly, Berkner and Shapley had a long history together because of overlaps in their radio propagation interests. Shapley also served on the URSI committee. He therefore had exposure to the international scientific community prior to the IGY as well. This array of pre-IGY connections helps explain the young scientific administrator's rapid and unanimous selection as the USNC vice chair. Stanley Ruttenberg, who served as secretary to many IGY scientific panels, later described Shapley as something of a "young Turk." Ruttenberg also described the younger man as "sort of like a Walt Roberts."

Despite Shapley's relatively junior scientific status, Ruttenberg viewed him as one of the prime forces in shaping and directing the IGY.[34] Many others did as well.[35]

In addition to the social scientific network as a factor in making Boulder an IGY hot spot, the Central Radio Lab's mission at the Boulder NBS labs centered on core IGY investigations. Shapley stood at the center of those activities. As Shapley put it in handwritten notes to a speech he gave on the IGY, "thus, a great deal of CRPL's work is by its very nature IGY work." He emphasized that as a place of global radio propagation studies, CRPL activities reflected the IGY approach because "an important fraction of CRPL's work from the very beginning has been of a type requiring collaboration and cooperative work with other U.S. laboratories and our overseas counterparts."[36]

Shapley understood that this boded well for the fortunes of his lab in Boulder and for the city itself as a place for IGY sun–earth science. His "Sun–Earth Interactions" division was the nexus of many CRPL solar–terrestrial research activities. It held this position because a primary mission of CRPL related to "the nature of the media through which radio waves transmitted" and how that media interacted with the waves.[37] Shapley, keenly aware of the unique circumstance presented by this major international effort, wrote to NBS head Alan Astin in early 1954 that "the IGY provides the Central Radio Lab with an unparalleled opportunity" and that "a large fraction of CRPL current activities have objectives quite similar to those of IGY."[38] He could have written the same sentence about Boulder.

IGY Science in Boulder

All the parts were in place to make Boulder a centerpiece of IGY research activity—a basic infrastructure based on sun–earth science and the personal associations that strongly connected Boulder to national and international IGY science. Understanding this, the relative ease by which Boulder transformed into a major site for IGY science is therefore not difficult to explain—of the twelve technical panels the USNC had formed by 1958, Boulder-based groups had research activities related to seven of the panels before the IGY began. These involved aurora and airglow, geomagnetism, cosmic rays, ionospheric physics, meteorology, solar activity, and "World Days and Communications" that coordinated IGY activities.[39] The majority of these activities resided in HAO, CRPL, or the University of Colorado Upper Air Lab. The scientific links between Boulder and the IGY effort, strong from the start, only intensified. By 1958, Roberts headed the solar panel to which the USNC appointed him, and Shapley, in addition to his vice-chair duties, assumed the chair of the important World Days and Communications panel. These links propelled Boulder further along its trajectory as a scientific research center after *Sputnik* in 1957 and the post-IGY years.

As the predicted maximum solar activity in 1957–58 fixed the date of the IGY, HAO's solar research and the sun–earth science facilities made it an inevitable center of IGY work. Roberts had a built-in tie to the very core of the IGY's *raison d'être*, and he understood this very clearly. Writing to Colorado senator Gordon Allott for his support of IGY appropriations, Roberts ventured that "Colorado research institutions are playing a big part" and that HAO had a role "at the heart of the whole IGY program."[40] Such funding eventually came from the U.S. government via the NSF.[41] As a result, Roberts could expect funding through the U.S. National Committee for his diverse IGY activities.

Because of HAO's extensive involvement in IGY science, a detailed list of its diverse activities in doing solar research is beyond the scope of this study. Suffice it to say that both the Climax site and Boulder offices of HAO performed important work in support of the IGY both as a research site and operational center for warning about solar activity. This work in turn brought more funds to the observatory. A few examples give the sense of IGY activities in which HAO participated.

At Climax the continuous observations of solar coronal phenomenon required some IGY support for the coronagraph flare-monitoring equipment. In Boulder, solar flare patrol using a standard telescope with hydrogen alpha light (656.6 nm) filters served as an important IGY observing tool in addition to the coronagraph at Climax. Both these observational techniques allowed much better characterizations of those eruptions near the sun's surface. Flares are the primary, although not exclusive, cause of disruptions in the earth's space environment. Many researchers across the world viewed the HAO data as vital to the IGY research agenda of better understanding how the sun affected the earth. Some of these researchers were just across town in Boulder at the Central Radio Lab.

HAO, as well as reporting its own data in real time, also became the central repository for the archiving of all IGY-obtained solar data.[42] These various data centers formed a crucial part of the IGY—researchers heavily involved in the data collection could not perform analysis in real time. Hence, these centers ensured the data would be stored and available for all time after the limited IGY period. Extensive data reduction and research eventually followed the real-time data collection, ensuring that HAO (and

Boulder) remained a center of IGY science long after December 31, 1958. These centers served as archives of IGY information, ensuring the safekeeping, reproduction, cataloging, and availability of collected data.[43]

Government agencies other than the NSF also had interests in IGY-related research, including the Air Force and ONR. From the long history of relationships with the Air Force and ONR, HAO had a built-in constituency that could augment IGY funding. The Department of Defense (DOD) would not directly fund IGY activities centrally except for logistics and research satellite programs, so it was up to individual services to support any scientific activities they saw useful.[44] IGY scientific support from DOD only came from service-specific research funding activities, such as ONR or the Air Force Cambridge Research Center, on a project-by-project basis. In many ways, DOD support for basic science just continued pre-IGY associations and activities.

The Air Force contributed to IGY in a number of ways through its continuing association with HAO. In addition to funding basic research and operations, the Air Force also agreed to support the development of HAO instruments for observing the total solar eclipse in October 1958.[45] The eclipse represented the sole IGY-sponsored opportunity to investigate this type of phenomenon. The event represented not only an important part of HAO's IGY science program, but also a unique opportunity in the history of astronomy, as many nations could participate. The cessation of this funding could (and did in 1957) cause problems for Roberts.

As Shapley noted, CRPL in effect served as something of a mini-IGY by its very nature. It therefore became a focus of intense IGY activity. A detailed discussion of the activities of this multifaceted organization during the IGY is beyond the scope of this work. A short review renders the spectrum of its contribution in making Boulder a scientific center during the IGY.

One example of CRPL's close relation to IGY began when Boulder scientists designed and constructed the latest version of a device for sounding the electrical properties of the upper atmosphere called an ionosonde. This instrument, based on radar principles, allows the investigation of the electron content versus height in the upper atmosphere. This work constituted a major part of CRPL's activities as it coordinated all ionosonde research around the world.[46] These activities dovetailed very naturally with HAO's solar work and observations, as the ionosphere tended to be the physical realm where HAO and CRPL interests really connected—an important link of the sun–earth connection. The ionosonde effort included the installation and supervision of these devices for the USNC in South America, in collaboration with international partners. These stations covered an important gap in knowledge about the equatorial ionosphere. They therefore contributed to a much better understanding of the earth's ionospheric dynamics and its effects on radio propagation.[47]

CRPL also designed and built airglow photometers for the U.S. IGY effort, including intercalibrating IGY airglow stations around the world. Airglow, a phenomenon whereby the night sky on occasion brightens, produced much curiosity among scientists. CRPL's efforts in coordinating the IGY airglow program enabled scientists to better quantify the phenomenon, although these efforts did not yield an immediate answer to the scientific puzzle.[48]

The Boulder group had many more involvements with the IGY. It supervised the U.S. program for studying significant radio propagation degradation that occurred during solar activity known as a "sudden ionospheric disturbance." The lab also planned a major part of the IGY effort to understand the nature and sources of atmospheric noise at radio wavelengths.[49] CRPL collaborated with the Air Force on the transionospheric phases of rocket flights, and after *Sputnik* they did experiments on the propagation of satellite signals through the ionosphere. In all of this sun–earth research, CRPL worked synergistically with HAO. This collaboration in many ways mimicked the relationship between and close association of Roberts and Shapley.

The global science of the IGY reflected itself heavily in Boulder science and stimulated the burgeoning astrogeophysical work that began there in the late 1940s and early 1950s. IGY activities required extensive coordination via the alerting of geophysically significant events that Boulder organizations could provide. The globally collected, disparate data sets needed repositories from which researchers anywhere could obtain these results, even years in the future. Roberts and Shapley worked to ensure that Boulder became this locus of "world warning" (of solar disturbances that might trigger geophysical responses) and IGY data storage.

Boulder as a Center for "World Warning" and the World Data Centers

Soon into IGY planning, the USNC established a technical panel with the nondescript title "World Days and Communications," with Shapley as the chair. This vague title masks today what was a crucially important IGY function—alerting the IGY scientific community to special geophysical events. This alerting function resulted from the need for special and highly coordinated global observations. Although not representing a scientific discipline, this panel provided a "scientific service," to use Shapley's words, for other

IGY disciplines and studies.[50] This work was an obvious extension of the CRPL prediction and warning efforts.

The CRPL had provided this alerting service since WWII, thus making the Boulder lab ideally suited to perform similar duties for the much-expanded IGY. As many of these special geophysical events and warning periods resulted from solar activity, CRPL established even stronger ties to HAO. Boulder therefore stood at the center of the IGY world for alerting participants worldwide about impending solar and geophysical activity. These warnings could notify the IGY community that a geophysical event, such as a solar flare, might occur (or had just occurred). These warnings ensured the comprehensive scientific observation of a solar or magnetospheric event and subsequent effects at the earth. "Perhaps the most important aspect of the IGY is the simultaneity of all the related scientific programs," Berkner wrote.[51] CRPL provided the link that produced this "simultaneity."

Shapley understood that this new task, as central to the IGY as it was, required much work. It also had significant payoffs for CRPL. "We are definitely saddled with this 'Special World Day' job," Shapley wrote to a friend in September 1954, speaking of the idea that would eventually result in the "Special World Interval" (SWI) warnings for IGY researchers.[52] Despite Shapley's comment about getting "saddled" with the alerting job, it made much sense that Shapley and CRPL found themselves at the center of this important activity for the IGY. Shapley himself summed up his views for Alan Astin, the director of the NBS, in early 1954: "the IGY provides CRPL with an unparalleled opportunity."[53] The "opportunity" he alluded to centered on making CRPL a world center for IGY sun–earth science, not just during IGY, but for years afterwards.

In 1942, IRPL had begun sending out predictions of ionosphere propagation parameters to the U.S. armed services. It soon relied on Roberts's coronagraph observations of the sun to form a basis of these forecasts. By 1949, in order to continue supplying the propagation forecasts, the CRPL established a site called the North Atlantic Radio Warning Service at Fort Belvoir, Virginia. NBS constructed the site before the move of the rest of CRPL to Boulder in 1954. The site not only provided radio propagation prediction services, but also collected and processed many types of radio, solar, ionospheric, and geomagnetic data derived from worldwide observations. Short-term warnings (for less than twelve hours in the future) of worldwide radio propagation conditions were the primary product of the warning service.[54] By 1951, a similar operation began under CRPL auspices in Anchorage, Alaska, called the North Pacific Radio Warning Service.

The continuation and extension of this CRPL work, aided by Roberts's HAO, into IGY made good sense to those involved in IGY planning. It was Berkner who originally proposed that CRPL take on this responsibility in July 1953.[55] Some French members of IGY who thought communications between the United States and Europe might be a problem questioned the selection. Any objections they had did not impact the final CSAGI decision to have CRPL perform these warning functions.[56]

The CRPL warning effort then became the focal point, or to use Shapley's words "the king pin," of the global IGY network of warning messages for events of solar–terrestrial import.[57] Their previous warning work had pre-positioned CRPL (and Boulder) to serve as an operational center of IGY. This work, along with Roberts's, also made Boulder a center of the intellectual underpinnings of not only IGY, but the extensive subsequent scientific activities resulting from the global effort.

Shapley oversaw the overall network, and CRPL's Fort Belvoir operation assumed the title IGY World Warning Agency (IGYWWA).[58] To facilitate these warnings, the concept for "World Days" and "Special World Intervals" arose within the CSAGI. During the World Days, specialized or more intense observations occurred over a given day or sequence of days. This category also included Solar Activity Alerts during periods of increased solar events. The latter, SWIs, occurred on short notice when forecasters predicted periods of auroral, ionospheric, or geomagnetic activity. Eventually, this warning work extended to predicting satellite parameters after *Sputnik* in October 1957. The relatively reliable and extensive communications network established for geophysical warning purposes also served usefully for more routine IGY activities as well. As IGYWWA made all decisions concerning these geophysical alerts, Boulder via the CRPL stood at the center of IGY daily operations.

The sun cooperated with IGY and produced high levels of activity—record breaking, in Roberts's account. The activity triggered many of these alerts.[59] The special observations generated by these warnings, along with the more routine observations, produced enormous amounts of data. The CSAGI understood early on in IGY planning this aspect of IGY research. This vast amount of data was "a common fund of knowledge open to all," as Chapman phrased it.[60] The question was what to do with this data that was too voluminous to analyze upon collection. Boulder, with Roberts and Shapley again taking the lead, found itself an integral part of the answer to this question.

As IGY was global in scope and international in execution, CSAGI hit upon the idea of three large international repositories of data. In theory, all researchers could access any data acquired by IGY.[61] This approach fit well into the open and free exchange of scientific results that characterized Boulder's developing sun–earth connection scientific community.

Considering locations that had the desire to perform this function and resources to do it adequately, CSAGI decided upon three World Data Centers (WDCs)—one each in the United States (WDC-A) and the Soviet Union (WDC-B), and another distributed among other, primarily European, countries willing to participate (WDC-C). Each WDC distributed specific discipline archives as it saw fit within its national borders, and one of the three WDCs would get the designation of "chief world data center" for that discipline.[62] As an example, HAO was the primary world data center (WDC-A) for solar coronal observations, with WDC-B (Moscow) and WDC-C (Pic du Midi) as secondary collections.[63]

Under each of these three national data centers there existed disciplinary data centers that dealt with specific data sets—solar activity, solar magnetic fields, ionosphere, and the like. Each WDC had a complete set of data, and the United States and USSR offered to finance all the centers and subcenters.[64] By agreeing to participate in IGY, all nations also agreed to send any data they collected to the WDCs.[65] The United States, USSR, and international WDCs assigned disciplinary subcenters internally to existing scientific and academic activities as they saw fit. All the WDC's ensured both the archival and continued access, at least in theory, of any data collected indefinitely after the formal end of IGY in 1958.

Boulder secured three of the IGY WDC-A subcenters, each coming by a different path. The WDC for solar activity came to Boulder via Roberts: HAO and the University of Colorado. The other two U.S. subcenters for ionosphere and airglow came via Shapley: CRPL and NBS. The preexisting activities that Roberts helped to establish proved crucial to the CSAGI and USNC decision to create these centers in Boulder.

Existing expertise stood as an important issue in deciding where to locate the WDCs. In the mid-1950s, Boulder had this expertise, as Roberts and the others had established the foundation of the horizontally integrated intellectual community based on the overarching concept of the sun–earth connection. WDC locations fit well into this community.

"The reasoning here is that the centers should not be purely archival," Shapley wrote to NBS Boulder Lab director Francis Brown. Rather "they should be attached to an institution which has thorough expertise in treating at least some of the data involved."[66] Shapley wrote this letter to Brown to apprise him of the possibility and importance of bringing a WDC unit to

Boulder. "Boulder is the natural site for one of these World Data Centers, in view of its growing stature as a place for geophysical research," Shapley added.[67] Roberts no doubt approved of his friend's analysis.

Ward Darley at the University of Colorado, most likely inspired by Roberts, led the first charge to get at least one WDC to Boulder. In December 1955, he wrote directly to CSAGI president Sydney Chapman about "the interest" of the University of Colorado in the study of data obtained during IGY.[68] Darley probably knew Chapman personally through Roberts. The letter did not have to travel far, for IGY chairman Chapman was in Boulder again visiting HAO and NBS.[69]

Geophysicist E. H. Vestine, director of the IGY's Data Coordination Office, responded to Darley in June of the next year. Vestine wrote that he was "gratified" that Walter Roberts invited him to the Institute for Solar Terrestrial Research's sun–weather symposium. The USNC had started a search for sites on June 1, 1957, to support the various subcenters for WDC-A. Vestine wished to determine if the university wanted to be considered as a location and to obtain the required NSF support. As Vestine planned to attend the ISTR symposium, he suggested meetings about the issue, including discussions with the Boulder NBS staff. Chapman, he thought, could participate as well "in the event he is about."[70]

Boulder, from the earliest stages of planning for these WDCs, ran ahead of many other sites (such as the solar facilities at Mount Wilson Observatory) as a potential location for at least some of the subdiscipline centers of WDC-A. Even a staunch supporter for Boulder-based IGY activities such as Shapley questioned some of these discussions between Darley and the IGY staff. He wrote to Joseph Kaplan in July that these meetings seem "to pre-judge the outcome" of the studies to locate the subdiscipline centers.[71] Shapley had a difficult role to play as both a high-level IGY official and a Boulder advocate. Here perhaps he tried to show his evenhandedness.[72]

The discussions proceeded apace over the summer and fall of 1956, regardless of any concerns over "prejudging." The efforts of HAO and CRPL to obtain a section of WDC-A soon converged. Proponents of bringing more IGY activity to Boulder joined forces to get not just one of the prized NSF-funded centers, but three of them—solar activity, ionosphere, and airglow. The last of these, a phenomenon of the upper atmosphere, results from solar electromagnetic radiation interacting with atmospheric gases and reveals an important aspect of sun–earth connection processes.

In July, Roberts wrote to E. O. Hulburt of USNC telling him of the recent discussions in June with Vestine in Boulder. Roberts also advocated again for Boulder as a site for the three archives. The primary point that Roberts wished to make centered on the new plan to make an HAO–CRPL joint repository—all those involved in these discussions agreed it was "feasible."[73] He and University of Colorado physics professor Wesley Brittin wrote to Darley earlier in June that NBS, HAO, and university representatives had "concluded the logical way for the Boulder scientific community to offer participation in this activity is for us to propose to establish a Boulder data subcenter" via this cooperative arrangement.[74] They also suggested that in addition to Boulder as a location for the three primary centers, it also made sense to make Boulder the secondary (duplicate) center for no less than four other disciplines—geomagnetism, cosmic rays, aurora, and rockets–satellites.[75]

Reminiscent of the CRPL move earlier in the decade, Roberts and the University of Colorado thought that offering a possible enticement to the USNC to put the new facilities in Boulder might help the decision along. "The present plan, contemplated by us in Boulder," Roberts wrote, "includes the provision of a suitable new building" on campus situated "closely adjacent to National Bureau of Standards laboratory." Vestine, he added, "is quite familiar with the facilities here, and the prospects for support by the Boulder scientific community."[76]

In September, Darley wrote to Vestine to state interest in housing a data center and a Data Coordination Office. Darley offered to construct a building near the new Sommers–Baush Observatory for this purpose. In addition to laying out the reasons he thought Boulder should get the centers, the university president suggested that $120,000 would cover the operating costs of the center. Darley obtained this figure from Roberts and Brittin's June memo where they laid out the situation for the university president.[77]

To follow up on the burst of activity related to the WDCs over the summer, Roberts suggested to Darley that he visit Vestine, Odishaw, and Berkner in October on a planned trip to Washington, D.C. Roberts wanted to discuss plans for the WDCs and the disciplinary subcenters with Berkner at an upcoming CSAGI (September 1956) meeting in Barcelona.[78]

USNC did not release its decision on the locations of the major subcenters until December 1956, but Roberts felt confident enough to write to a colleague in late November that three of the IGY disciplines "will have the primary data resources here in Boulder."[79] After all the lobbying of the existing sun–earth community, the USNC decision in December that Boulder was to get the three primary data centers of solar activity, ionosphere, and airglow appeared anticlimactic.[80] Yet, this decision represented quite a coup that further established Boulder as city of atmospheric and space science—

obtaining even one of these centers made a location an important nexus of global science.

As the collectors, adjudicators, and permanent repositories of the vast amounts of data obtained during IGY, these centers stood as key sites of disciplinary activity both during IGY and after. The recognition of Boulder by the USNC represented a milestone in the development of the city's scientific reputation, given the criteria by which the USNC adjudicated the locations. Only the Washington, D.C., area found itself home to more of these important data centers (four) than Boulder. These two cities were the only U.S. locations to acquire more than one center. Given that less than a decade before Boulder had the reputation of a "scientific Siberia," the award of these highly visible IGY data centers served as an important indication of the city's rapid rise to prominence as a place for scientific research. These centers came to Boulder not randomly but precisely because, as Roberts and others understood, the city was well-poised to house them by the mid-1950s. By the efforts of Roberts and others in the preceding decade, Boulder achieved a reputation as an important site for sun–earth connection science. As the city's reputation as a science center grew, so did its possibilities.

The existing sun–earth social community, including Roberts's close friendship with Chapman, played a significant role in this outcome. Shapley's dual role as IGY official and Boulder advocate also played a crucial part in bringing IGY science to the city. HAO and CRPL also had the expertise in the relevant disciplines and handling the requisite data sets, an important criterion for a data archival site. In addition, CRPL was an obvious choice as the IGY center for world warning. Opportunity, preparedness, and ambition joined to give Boulder this important addition to its rapidly growing science community.

Even with all this activity generated by IGY and Boulder's role in it, HAO still struggled with funding issues. After some serious setbacks in 1957, however, new and unexpected events helped to elevate Boulder's prominence in the U.S. science.

"The Thing that Really Saved Us Was *Sputnik I*"

The diverse scientific activity of Roberts and the HAO in the mid-1950s masked a serious and systemic problem. Despite the successful fundraising efforts; continuous government funding via ONR, the Air Force Cambridge Research Center, and CRPL; and the IGY-funded work, the early and mid-1950s saw the HAO struggling financially as a viable scientific organization.

"We were having some hard times," Roberts recalled, reflecting upon HAO's precarious financial state in the mid-1950s. He even had difficulty meeting staff payrolls.[81] The diverse scientific activities of Roberts's mini solar empire in Boulder had significant amounts of overhead cost that the diverse funding sources only partially compensated for. As the HAO seemed always on the edge financially, all the diverse sources of Roberts's funding were vital in keeping the HAO running.

John Jeffries, another noted solar astronomer who had spent part of his early career in Boulder, recounts that Roberts in the summer of 1957 "came to see me one day . . . to ask whether Charmain and I could survive for a month without pay." This request surprised the young astronomer since "Walt made sure that we didn't worry about anything but our science."[82] Decades later, Roberts could joke about these periods, saying that the pay at HAO sometimes consisted of "paperclips or something like that."[83] This latest episode in Roberts's financial woes resulted in part from a funding reduction in that year. The planned termination of Roberts's Air Force funding in 1957 precipitated this latest crisis for HAO.

Charles E. "Engine Charlie" Wilson, President Eisenhower's Secretary of Defense, had no love of basic scientific research. He once commented that "basic research is when you don't know what you are doing."[84] Although the president had a higher view of science and scientists than his defense secretary did, the parsimonious Eisenhower nevertheless directed the DOD to reduce research funds.[85] Wilson then triggered the research funding crisis among the nation's scientific community in the summer of 1957, about the time of the IGY start, when he ordered a 10% reduction in support of military-sponsored basic science in a move to lower DOD expenditures.[86]

Roberts received word by telephone from former HAO colleague Jack Evans at Sac Peak, on August 28, 1957, that the Air Force would suspend HAO research contracts in the near future.[87] Once begun, this cessation of funds could last, Evans reported, until at least the beginning of 1958. Roberts thought that "the situation looks grave," as he told the HAO supervisory committee in early September.[88] The two contracts affected struck at the core of HAO and IGY scientific activities. One contract included IGY support for solar observation. The other contract provided for the development and construction of instruments to observe, as part of the IGY effort, a total solar eclipse in the South Pacific on October 12, 1958.[89]

Evans had to pass on more bad news to his friend Roberts. The contracts selected for termination would also suffer a 5% reduction until funding stopped, and the Air Force contracts not affected (because they had end dates later in 1958, apparently) would also suffer from the same reduction. Finally, the Air Force directed the spending rate for the contracts. In addition to the funding cessation that he thought "utterly irrational," Robert Low, HAO's business manager, felt particularly incensed by the micromanagement of the spending rate. He commented to the HAO trustees in a memo that "we can't conduct a sound research program with such a limitation of flexibility."[90]

The gravity of the Air Force move required immediate action on the part of Roberts and others affected if there was any hope in getting the decision reversed. Within days, Roberts enlisted a small army of administrators and science policy-makers in an attempt to address this crucial problem with potential effects on the IGY program. Among those he spoke to first on the matter were USNC's Hugh Odishaw, Lloyd Berkner, and new University of Colorado president Quigg Newton.[91] Roberts informed Naval Research Laboratory solar scientist John P. Hagen that there was "lots of push getting started" to avoid the funding cutoff.[92]

Roberts soon drafted a letter for Quigg Newton to send to Colorado's U.S. congressional representatives. Former politician Newton was a strong advocate, like his predecessor Ward Darley, for making his university a place for research science and was therefore anxious to help reverse the cuts. The funding freeze, Newton wrote, "will deal a crippling blow to our work and to basic research goals of the Air Force if it is not dealt with promptly." He added that "the crisis is intensified because the Observatory has been making special efforts, with Air Force encouragement, to accelerate this work to meet deadlines" for IGY. He expressed concern about the resulting lack of support for IGY, a U.S. "prestige loss" among other nations, and the loss of important scientific personnel from HAO and Boulder. The university president added, "I know you appreciate the importance of the University research program to Colorado, and therefore I have not hesitated to call on your help in this emergency."[93]

Only a day after Newton drafted his letter, Hugh Odishaw wrote to Roberts and Earl Droessler, the Executive Secretary of the Coordinating Committee on General Science in the DOD's Office of Assistant Secretary of Defense for Research and Engineering (OASDRE). In the letter to Roberts, Odishaw commented that "the situation is widespread and looks very serious," and that he had already called Droessler.[94]

In writing to the DOD official, Odishaw pointed out that HAO difficulties and DOD reductions in basic research funding "now appear to be seriously jeopardizing large portions of the IGY program." He emphasized the point that "disastrous scientific consequences" would follow unless the DOD restored the funding. Among his concerns were "international complications" and a possible "major disruption" to the U.S. IGY effort.[95]

As he had for the better part of the decade, Roberts scrambled for funds. In an effort to seek funding to cover the Air Force losses, he submitted a proposal to Odishaw and the USNC. His idea entailed working on "important data analysis problems in the solar-geophysical areas," for the Air Force funding cut had made some manpower "unexpectedly available."[96]

The widespread, high-level, and intensely unfavorable response to DOD funding cuts in general, and HAO cuts in particular, did appear to generate some positive response from the DOD rather quickly. By mid-September, President Newton had spoken to Richard Horner, Assistant Secretary of the Air Force for Research and Development. The official indicated the contracts picked for termination would not be randomly chosen after all, but based on a selection process. HAO and the University of Colorado had some hope.[97]

By the first day of October 1957, the administrative officer of the USNC, B. Van Matter, wrote Roberts. He reported that Odishaw had discussed the crisis with DOD officials. Odishaw "had been assured that they are doing everything possible" to continue support of basic research. Van Matter added that, not realizing the Soviets stood on the verge of launching *Sputnik* a mere three days later, the DOD had already created "some measures of relief" for IGY-related work. These measures, Van Matter wrote, "should be effective within a few weeks."[98]

The real relief ultimately came not from the Pentagon, but from a most unexpected source—the Soviet Union. The launch of *Sputnik* on October 4 rather quickly and permanently changed the landscape of American science.

The many opportunities afforded to Roberts and the Boulder sun–earth connection community by the IGY paled in comparison to those presented in subsequent years by the altered U.S. policies for space and atmospheric science. After *Sputnik*, according to an HAO scientist, "Walt, with HAO, never looked back."[99] In many ways, the Soviet launch not only saved the HAO program in Boulder, but it also ensured the continuity of Boulder's growing complex of sun–earth science activities. The reaction to *Sputnik* also set the stage for further growth in Boulder-based space and atmospheric science over the next half-decade. As Roberts summed up the situation most accurately years later, "the Russians bailed us out."[100] What happened in Boulder after the Russians inadvertently rescued HAO follows.

CHAPTER SEVEN

An Atmosphere of Change

"There Is a Tide in the Affairs of Men"

A new pattern for academic science arose quickly in the U.S. response to *Sputnik*. Prior to its launch, a struggle developed in U.S. science policy. Some thought federal research funds should primarily go to agency-sponsored programmatic research. Others wanted more funds for basic scientific research guided by universities and determined by their scientific faculties. An "immediate effect" of *Sputnik*, one historian writes, "was to resolve this issue unequivocally for the university position."[1]

President Quigg Newton of the University of Colorado was Roberts's close associate, as were his immediate predecessors, and tied into Boulder's sun–earth science community. Newton understood that this intense U.S. reaction to *Sputnik* and its attendant "new pattern" for academic science afforded opportunities to the national higher education system—his university in particular. Newton, the former Democratic mayor of nearby Denver, served as a vice-president of the Ford Foundation before his selection as the university's eighth president. As he assumed his new role in December 1956, Newton had the desire to make a university that ranked "among best in the world."[2]

Yet, even his provost, Oswald Tippo, understood the university needed improvement as the school was "still a long way from the top."[3] Newton's vision did not differ significantly from his predecessors Stearns and Darley. But the new president had the great good fortune to benefit immediately from the restructuring of U.S. science policy and resulting increases in federal support for academic science in the late 1950s.

"There is a tide in the affairs of men," Newton often liked to quote from Shakespeare's *Julius Caesar*, "which, taken at the flood, leads on to fortune."[4] Newton understood that "there were many factors in the university's favor at that particular time" and that this unique combination of factors might "not repeat themselves for a long time to come, if ever."[5] He was correct in this observation—this was a watershed moment for American universities. President Eisenhower said that his "scientific advisors" warned him that the country (by its universities) needed to produce scientists, "thousands more" by their reckoning.[6] Congress showed even more enthusiasm than the president had about entering a techno-cultural race with the Soviets. Money going to American institutions of learning therefore quickly grew to a "flood stage," coming from "every tributary" and going to academic research.[7] Newton did not stand alone in appreciating the opportunity that *Sputnik* afforded U.S. science and academe. Many understood—scientists, university administrators, even some in the local populace surrounding universities—that a unique period had begun for American science in the wake of *Sputnik*. "Bountiful times" were about to come to all sciences, and space sciences in particular.[8] Between 1959 and 1964, for example, federal funds for university research increased by almost 200%.[9] In this heady era, the National Science Foundation surpassed even the DOD as a sponsor of academic science.[10]

As science and education became central arenas for this new type of international competition, Walter Roberts found fundraising activities unnecessary.[11] Boulder stood in many ways uniquely positioned to contribute to this new "competition" primarily fueled by government money. The intellectual community based on the science of the sun–earth connection had thrived during the early and mid-1950s and only grew more given the burgeoning of U.S. science in the late 1950s.

Roberts, closely working with Newton and others in Colorado, aggressively sought these diverse new opportunities to further propel the university and Boulder into the first rank of atmospheric and space science research locations. Another opportunity arrived when the National Academy of Sciences decided that the time had come for the United States to have a world-class national institute for atmospheric research. This possibility was yet another opportunity for Boulder to establish itself atop the pinnacle of places in the world to perform atmospheric and space science research. How Roberts and others in Boulder took advantage of this rare opportunity reveals much about the working of U.S. science policy decision-making in the late 1950s.

Lécuyer discusses how "competencies"—specific expertise— in engineering, management, and production help to explain the rise of Silicon Valley.[12] A similar situation occurred in Boulder. In Boulder's case, the competency or expertise established centered not on the manufacturing of a product, but on the production of knowledge relating to the physical understanding of the sun–earth connection. By the late 1950s, Boulder had achieved an international reputation as a place for these and associated studies such as atmospheric science.

Roberts had helped situate these competencies in Boulder. His increasingly deeper involvement in the sun–weather dimension of the sun–earth connection provided a logical bridge from solar physics to the study of the earth's atmosphere. The coupling of these two regions physically via the sun–earth connection mandated a better understanding of the effect of solar activity on the earth's upper atmosphere. His interests in atmospheric science had their roots first in his intellectual curiosity and then developed into a practical device for fundraising. As his interests expanded, so did his influence, not only in Boulder, but in national scientific circles as well. AnnaLee Saxenian highlighted the importance of a community of experts in the rise of Silicon Valley.[13] Similarly, the scientific community based on sun–earth connection studies that Roberts helped to create, both in Boulder and beyond, enabled the city to acquire even more status as a city of scientific knowledge production.

In this fertile atmosphere for scientific research in the late 1950s and early 1960s, Roberts, the astronomer, in just a few years found himself at the very top level of policy-makers and administrators in U.S. atmospheric science—an unexpected development. Roberts lacked any formal training in atmospheric science and did no research in the discipline except for his controversial studies in the possible connection between solar activity and earth's weather.

As a scientific policy-maker, however, he was in a key position to bring more sun–earth science to Boulder. This post-*Sputnik* phase of Boulder's growth had two fundamental and interrelated reasons behind it, both with origins earlier in the decade. First, there arose earlier in the 1950s increasingly widespread concerns by leading practitioners in the discipline about the intellectual state of U.S. meteorology. Second, high-level U.S. policy-makers and scientists expressed concerns about intentional weather modification (often called "weather control"). These two sets of concerns eventually translated into funding decisions that brought more science to Boulder.

This and the following chapter analyze the complex series of changes in U.S. science policy in the pre- and post-*Sputnik* era that put Boulder on the top rung of U.S. science locations. In this chapter, beginning with NAS and NSF, the focus of discussion is on how U.S. federal policy for atmospheric science adapted to the perceived challenges of the early Cold War era and how Walter Roberts participated in these events.

The Restructuring of U.S. Science Policy during the IGY Era

The U.S. reaction to *Sputnik* brought an immediate and stunning reversal of fortune for Roberts and HAO. Only a few days after *Sputnik*, the new Secretary of Defense, Neil McElroy, overrode Wilson's cuts and reaffirmed the role of basic science to national security.[14] A "honeymoon fervor immediately returned to the romance between science and government," wrote one long-time observer of U.S. science policy.[15] Given the emotional, and in some cases panicked, national reaction to *Sputnik*, this metaphor seems appropriate. "All hell did break loose," one scholar of the period wrote.[16] Honeymoon or hell, this was a period of further growth for Boulder as a science center.

Short-term effects on Boulder's development went far beyond just the restored DOD funding. The IGY immediately enlisted the HAO to provide data on earth's new artificial companion, for the earth satellite program was an essential part of the IGY program.[17] In November, Roberts and two members of his staff, John Warwick and M. Bretz, sent a proposal to USNC for a daily analysis of the satellite's position and ionospheric data obtained by radio signals emanating from the spacecraft.[18]

Roberts clearly understood this swift change in fortune. He wrote to a potential private donor on November 1, 1957, that with respect to funding, *Sputnik* "seems to have improved the picture substantially," and, therefore, "I think there is now clear sailing ahead."[19] For Roberts, in addition to a revived HAO, the new landscape of U.S. science presented an important and interesting twist. His professional future no longer centered on solar or space physics research as it had for almost two decades. Rather, the great new opportunities for Roberts presented themselves in the atmospheric sciences. This jump to a fundamentally different, and rapidly developing, discipline had consequences not only for the remainder of Roberts's career, but for the development of Boulder as a city of knowledge as well.

Before this happened, the shock of *Sputnik* had to reverberate around Washington, D.C., and then get translated into a strengthened emphasis for U.S. science policy-making. In response to the shock, new oversight, funding, and advisory bodies sprouted in the nation's capital.[20] "Nobody has—or could—come up with a readily comprehensible table of organization to explain the labyrinth of [scientific] agencies, foundations, consultant-ships, academies, and committees that has grown up in Washington in recent years," wrote an observer of the Washington policy scene in 1963.[21] These changes in the advisory and oversight structure of U.S. science resulted in new scientific (and associated academic) research polices for the nation in the physical sciences. This was particularly true for atmospheric science.

In addition to the National Academy of Sciences, the National Science Foundation came to play an increasingly important role in Boulder's fortunes. Congress created the NSF in 1950 after a long and intense struggle over the precise way in which it would function. Some argued for an organization of a closed circle of scientists and administrators making policy decisions. Other wished for a more open, democratic approach that enlisted advice from a broad base around the nation.[22] The result was something of a compromise between the two positions. Both the NSF's role in funding the U.S. IGY program and the U.S. reaction to *Sputnik* increased the agency's significance in overseeing U.S. science, a situation that Alan Waterman, the NSF director, eagerly exploited.[23] The NSF engaged on a new front in the Cold War. As many have observed, after *Sputnik* the U.S.–Soviet contest shifted to a more encompassing cultural and domestic competition between East and West.[24] Many benefited from this "permacrisis" that followed *Sputnik*, including American science and academe.[25]

Roberts received increased funding from non-DOD sponsors within the U.S. government, as did many other like-minded U.S. physical scientists of the mid-1950s. More money and a funding source possibly more stable than the DOD arose via an enhanced NSF. The agency's funding for basic science (not including IGY appropriations) increased over 300 percent between 1957 and 1959 from $40 million to $130 million. This increase came on the heels of previous increases in NSF funding from about $12 million to $40 million. The NSF budget therefore increased by more than 1000% between 1955 and 1959.[26] *Sputnik* accelerated a trend that had started earlier in the decade. Among the many effects of this new funding source, Roberts no longer had to undertake the arduous work of finding private sponsors for his work.

American Atmospheric Science circa 1950s

Detlev Bronk, president of the prestigious National Academy of Sciences National Research Council (NRC), created the Academy's Committee on Meteorology (COM) in 1956. He did this after Louis S. Rothschild, Undersecretary of Commerce for Transportation, asked him for assistance in helping the U.S. Weather Bureau (USWB) "vitalize" the science.[27] Although created prior to *Sputnik*, COM found itself energized by IGY and the events following the launch of the first artificial satellite. This committee also had an important effect on Boulder's development as a city of atmospheric science knowledge and Roberts's career as a scientific administrator. Why did Undersecretary Rothschild feel the need to make this request of Bronk?

Atmospheric science, maturing a great deal as a scientific discipline in WWII, still had significant issues facing it in the early and mid-1950s. Some historians note that meteorology had established itself as a serious academic discipline by 1950, especially by its use in WWII. Harper correctly points out that meteorology established itself as a true scientific discipline by the demands of WWII. But the strong operational and forecasting focus of the war years left significant gaps in its development as a serious, cutting-edge scientific discipline. Concerns about filling these gaps drove much atmospheric science development in the 1950s.[28]

As a result of these perceived gaps, there still existed in the early and mid-1950s a widespread feeling among atmospheric scientists that the discipline needed to develop further. They had concerns that their field was not yet a fully developed modern scientific discipline on par with other physics-based sciences, such as nuclear research or astrophysics. Among problems often cited by those concerned were the development of significant theoretically sophisticated research problems and methods. Atmospheric modeling using advanced mathematical techniques was still only in its infancy in the early 1950s.[29] Another, but related, concern expressed frequently centered on the lack of adequately trained personnel going into the science—many thought the best students went into more traditional physics precisely because of the perceived theoretical backwardness of the discipline.[30] Much of this angst focused on, and resided in, the U.S. Weather Bureau. This resulted in Rothschild's plea to Bronk for help.

Although not formally a part of the U.S. IGY effort, COM shared with USNC an institutional home at the National Academy of Sciences. A number of its original members had played important roles either directly or indirectly in IGY. Not the least of this group was the ubiquitous and influential U.S. entrepreneur of science, Lloyd Berkner. Bronk, somewhat tongue-in-cheek, wrote on his friend Berkner's invitation letter that he hoped to get wisdom from him "that flows from you like manna from heaven."[31] In addition to Berkner, Harry Wexler, the U.S. Weather Bureau chief scientist and its representative to COM, had significant roles in both the planning and execution of IGY. These connections further established high-level ties between IGY and U.S. atmospheric science.

Berkner professed no expertise in atmospheric science, as was true of a number of others in COM such as engineer Hugh Dryden of the National Advisory Committee on Aeronautics, influential mathematician John von Neumann, and outspoken physicist Edward Teller. Bronk picked them precisely because they were noted in their respective fields and influential voices of the U.S. scientific establishment. In doing this he wanted to provide outside stimulus to U.S. atmospheric science. He wanted "cross-fertilization" with other scientific fields to help atmospheric science with its apparent disciplinary woes.[32] All shared in the widespread opinion that in many ways American meteorology was theoretically lacking. As a result, these scientific heavyweights were willing to participate in the committee's deliberations and help provide this "cross-fertilization."

This desire for an interdisciplinary approach to invigorate the atmospheric sciences dovetailed nicely with solar astronomer Roberts's views on the best way to tackle the questions of the sun–earth connection. This confluence of Roberts's growing interests in the atmospheric aspect of sun–earth science and the strong desires of those in meteorology to reach beyond the discipline helps explain how non-meteorologist Roberts became a major voice in the development of U.S. atmospheric science.

Berkner was an outsider to the world of atmospheric science and meteorology, for his research focused on the extreme upper reaches of the earth's atmosphere. He only dealt with the lower atmosphere indirectly via his interest in radio propagation. He therefore had little scientific overlap with the bulk of atmospheric scientists. These researchers mostly thought about much lower regions of earth's atmosphere (and how to forecast weather). But, as was the case with Roberts, his strong interdisciplinary interests and entrepreneurial skills enabled him to play a major role in the development of atmospheric science.

Berkner accepted not only membership but also, "with surprise," the chairmanship of the committee. He stated upon acceptance that, as he was not a meteorologist, his freedom of action was "uninhibited."[33] Carl Rossby, one of the creators of modern American atmospheric science and a gifted student of Norwegian atmospheric science pioneer Vilhelm Bjerknes, served as co-chair. Rossby's presence ensured there was a strong meteorological presence on the diverse committee from the beginning. The group took a broad view of their charter to advise USWB. They plunged into issues well beyond immediate concerns of their weather bureau sponsors.[34] Rather than simply focus on the limited USWB concerns on how to forecast weather better and improve their staff's scientific abilities, COM wanted to investigate the entire state of the discipline, including the widespread perceived woes of the field discussed previously. As the NAS committee set about its deliberations, there was another important element of U.S. science policy entering into the world of U.S. atmospheric science occurring across town in Washington.

The NSF and the Control of Weather

The National Science Foundation was the other D.C.-based organization that affected Boulder's development. This occurred in the late 1950s when the agency decided to fund not just atmospheric research, but the study of weather control itself. This interest led to NSF funding HAO with small amounts of support in the earlier 1950s, including cosponsoring (with the American Meteorological Society) the summer seminar on sun–weather research at Roberts's Institute for Solar Terrestrial Research in 1956.[35] The true significance of NSF funding for Roberts did not come via support of HAO directly. Rather, it came from the NSF's decision to get into the atmospheric science world in a major way via weather control research.

The impetus for this action came as a result of the "Orville report," generated in late 1957 by a congressionally directed study of the possibility of weather control and climate modification beginning in 1953. This led to the increased NSF effort in atmospheric sciences. The move was part of the foundation's overall attempt to transition from the funding of small-scale efforts in the early 1950s to expanding into bigger science later in the decade.[36] This important turn of NSF toward the development of U.S. atmospheric science is not featured in some of the standard histories of the foundation, despite the importance of this move for both U.S. atmospheric science and the foundation itself. Perhaps NSF's other work in the late 1950s, for example, its efforts on the large national astronomical observatories, overshadowed its support of the development of U.S. meteorology.[37]

Weather modification ranked as an important national concern in the postwar era because of its perceived implications for the nation's security and commerce.[38] Some argued that weather, if controllable, could serve as a weapon during hostilities.[39] Those with interests in agriculture thought weather modification was a possible solution to the age-old problem of inconsistent rains and occasional droughts. Given this broad interest, there were many in the postwar years that thought weather modification a desirable thing to achieve. There were also a few scientists who claimed to know how to achieve it (at least to some degree).

In the late 1940s and early 1950s, a number of researchers engaged in radically different scientific approaches to the problem of weather modification. Among them included Nobel Laureate physicist Irving Langmuir of General Electric and high-profile private weather consultant Irving Krick.[40] Both these men began making extravagant claims relating to the efficacy of their methods on forecasting and rainmaking.[41] These claims and counterclaims pitted many involved against one another. All of the proponents pitted themselves against a dubious and seemingly hapless U.S. Weather Bureau that could not confirm (or disprove) these claims to the American public. Irrespective of the weather control controversy, personnel at the USWB mainly wanted to concentrate on forecasting the weather more accurately.

The weather modification controversy reached a fever pitch in the nation. Richard D. Searles, Under Secretary of the Interior, even accused the U.S. Weather Bureau of "ineptitude" in addressing this whole "rainmaking" issue produced in the nation by these claims and counterclaims of weather modification proponents in 1952.[42] Strong contention over methods, advocates versus skeptics, and legal issues arising from perceived damages resulting from the weather modification attempts served to create turmoil in U.S. meteorology. The USWB bore the brunt of the outcry. Spurred on by this highly visible controversy, and interested in the potential applications of weather modification both for agriculture and as a weapon, Senator Francis Case of South Dakota decided to do something about the situation. Case, apparently a weather buff himself, created and worked to pass a bill in 1953 that established a high-level committee to study issues of weather control and the general state of U.S. meteorology. Called the Advisory Committee on Weather Control (ACWC), officials selected meteorologist Howard T. Orville as its chairmanship. ACWC submitted its summary document, "the Orville report," to Congress in December 1957 and concluded its work.

Among the conclusions rendered in the report, ACWC suggested that the nation needed to do much more fundamental atmospheric research in order to better understand the possibilities, if any, of weather control. Another key recommendation related to the need for the production of more research meteorologists and atmospheric scientists. Both these conclusions resonated with the COM report, then not yet finalized. Perhaps the most consequential of the recommendations for all involved related to NSF. The report suggested NSF assume leadership in providing resources to accomplish these recommendations. This recommendation in essence put NSF in charge of coordinating U.S. weather control efforts, including broad areas of basic atmospheric science research.[43]

The Move toward a National Institute for Atmospheric Research

The exact manner through which NSF entered atmospheric science originated at the National Academy of Sciences. The NAS Committee on Meteorology, just about the same time as the release of the Orville report, produced its own summary of the state of U.S. meteorology. Released by the NAS in February 1958, the "Interim Report of the National Academy of Sciences Committee on Meteorology" made specific recommendations on increased funding and manpower requirements. The report also outlined the urgent need for a National Institute of Atmospheric Research (NIAR) to assist in the vitalization of atmospheric science. The NIAR's primary purpose was to augment the research done by university atmospheric science departments.[44]

COM acted to advance the recommendations of its own report. Committee members, motivated by internal disciplinary concerns, focused on the need to encourage more (and better) research scientists into meteorology. Other motivations included external factors such as a the fear that the USSR had developed a much better weather forecasting ability than the United States, creating a "forecast gap."[45] This "gap" was the meteorological equivalent of the other perceived gaps in the American consciousness engendered by *Sputnik* in the United States—missile, space, education, etc.[46] The real issue driving this gap phobia was the widespread idea that America had to be the world's leader in everything scientific.[47] In yet another scientific area, it appeared the U.S. lagged behind the Soviet Union—ominously so, as better weather forecasting might lead to better crop yields, military operations, and the like. As with other gaps, many in the nation expressed the opinion that the situation needed immediate high-level national attention. Atmospheric scientists and others involved with the COM eagerly gave it this attention.

COM met on January 31, 1958, in Berkner's offices in New York City with members of the wider community of atmospheric science, including universities and USWB. Brookhaven's lab director, Leland Hayworth, gave a presentation on the management structure of AUI and Brookhaven to the group.[48] Most likely Berkner, president of AUI, which ran the lab, arranged for him to speak to the atmospheric scientists.

Apparently taken by the momentum of events and with the AUI/Brookhaven model for a new national institute for atmospheric studies, the university meteorologists convened again at UCLA only a few weeks later. The primary reason for the meeting focused on the creation of an AUI-like arrangement for the proposed new national institute (NIAR). They agreed to first form a new University Committee on Atmospheric Research (UCAR) patterned after the AUI arrangement that was distinct from COM. MIT meteorologist Henry Houghton served as UCAR's first chairman. Although interested in advancing all of the interim report's recommendations, making progress on creating an NIAR stood prominent, for it served as a key in effecting the rest of the recommendations. The next question was who might fund such an institute.

As an advisory board, the National Academy of Sciences did not have funds to even begin implementing its COM recommendations. NSF figured prominently as an obvious candidate for funding this new scientific research center. According to one participant, Berkner "fingered" NSF as the organization best suited to advance the cause of atmospheric sciences in the manner that the committee report outlined.[49]

NSF support showed promise for UCAR in its effort to create an institute. Riding a tide of increasing funds, the agency appeared to have rid itself, at least in the short term, of the unstable funding that had greatly concerned U.S. physical scientists such as Roberts.[50] As more funds became available, the agency increased its portfolio of research it wished to sponsor, including, for example, significant work in shaping the U.S. biological research programs.[51] NSF director Alan Waterman's ambitious goals also included the funding of large-scale scientific research centers such as new national astronomical observatories.[52] NSF plans therefore fit well into UCAR's vision of a new national center for atmospheric science. NSF's philosophy of funding ideas that came mostly from the scientists themselves also appealed to UCAR members, as it did to most all in the scientific community.[53]

In March, Houghton submitted to NSF a modest $52,400 proposal—seed money for continued planning—to support UCAR's grander goals.[54] He had to submit the proposal under MIT auspices as UCAR did not yet have any legal status. It still existed only as an informal agreement among the university participants.

These independent review committee reports—one from the Orville group (ACWC) focused on weather control and the other written by COM on primarily disciplinary issues—produced complimentary conclusions that converged in the NSF. The agency was the logical choice for UCAR to approach for continued funding of the ambitious programs arising from the two reports. Fortunately, Waterman was flush with new funds in the wake of *Sputnik*. He became enthused about an entirely new discipline NSF could advance.[55] The NSF director turned a receptive ear to the two committees' recommendations. Even before ACWC delivered its final report, he aided Orville and Senator Case in creating Senate Bill S. 86 that officially transferred the Orville committee's work to NSF in 1958.[56] Waterman decided to make NSF into an overseer of U.S. atmospheric science and weather control—the agency played an integral role in the development of "big meteorology" in the United States. Boulder was an integral part of this development.

The Road to NIAR, then NCAR (and Boulder)

As UCAR continued its work implementing the COM interim report recommendations, President Eisenhower signed the bill getting NSF into weather research on July 11, 1958. The White House press release stated the act directed NSF to provide for weather modification research in all of its dimensions. In a rare instance of humor in a White House press release, the document commented on the bill as an attempt to address Mark Twain's lament that everyone talks about the weather, but no one does anything about it.[57]

More importantly, to help implement the act, Waterman created an atmospheric science program in the Foundation's Division of Mathematical, Physical, and Engineering Sciences after the bill's signing. This event was part of the pattern of NSF's burgeoning portfolio during this post-*Sputnik* era. A large beneficiary of this growth was earth science. Soon after the creation of the atmospheric science program, for example, Waterman created an oceanography program with the division as well.[58]

Earl Droessler, who had endured the U.S. scientific community's outcry arising from DOD basic research science cuts in 1957, relocated from the Pentagon to head this new, ambitious, and highly visible NSF effort.[59] With this announcement, the path to an NIAR (and eventually the renamed, NCAR) of some sort opened. All the pieces for this happened to fall in place. Most notable among them were potential funding from the government via NSF and an ambitious set of UCAR member universities to create this center.

Privy to all the events relating to the Orville report, COM, and defense meteorological funding efforts during the previous few years, Droessler came to NSF as an enthusiastic supporter of the proposed research center for atmospheric science. He immediately built funding for the effort into NSF budget requests for future years. Droessler did this even before NSF's overarching advisory body, the National Science Board (NSB), had an opportunity to comment on these new NSF initiatives.[60]

Waterman's actions sparked concern in and out of NSF. Dixie Lee Ray, a special assistant to the NSF associate director for administration, presented a very critical picture in a 1963 review of the process by which the National Center for Atmospheric Research (NCAR) became a permanent NSF ward. In a nearly one-hundred-page report, she argued that Earl Droessler made the decision to back NCAR as soon as he assumed his NSF duties. He then pushed the idea through NSF bureaucracy, in particular the National Science Board that oversaw NSF activities. It seems, as related subsequently, that Waterman did not need much pushing from Droessler or anyone else.[61] Even some in the atmospheric science community had reservations about the process undertaken for NCAR's creation.

Meteorologist Joanne Malkus spoke for herself and Columbus O. Iselin, noted marine scientist and director of the Woods Hole Oceanographic Institute, at a UCAR meeting. She expressed chagrin over the whole process, stating that the new center should "grow naturally" and that "*because of the small size of the field of meteorology, too many decisions have been made by too few persons*" (emphasis in the original).[62] Writing before NSF created its oceanography program, Iselin also expressed concern later in a letter to COM that NSF had been "the least liberal and understanding" of government agencies in supporting earth science research with "long range funds" and "adventure funds."[63] As a result, he declined the invitation of Woods Hole membership in UCAR.

Despite these concerns over process, the need for improved understanding of the atmosphere did strike a nerve in the nation. Even the U.S. Governors' Conference, at their meeting in May 1958, decided to get involved in the establishment of a new institute. The attendees passed a resolution supporting NSF's efforts in meteorology and cited the need for some sort of national atmospheric study center.[64] Given this widespread concern of the nation's meteorological health, the march toward a national site of atmospheric research continued. At this point, in the summer of 1958, the primary question UCAR faced centered not on the "if" of the institute, but rather what shape it might take with respect to organization and purpose.

UCAR's first serious step in answering this question came in February 1959 with its report "Preliminary Plans for a National Institute for Atmospheric Research—Second Progress Report of the University Committee on Atmospheric Research."[65] This unwieldy title soon gave way to the nickname "the Blue Book" because of the document's original dark blue cover. The work, authored by a small team under the direction of Traveler's Insurance Company meteorologist Thomas Malone, resulted from a series of 17 two-day meetings held mostly over the latter part of 1958.

Events progressed rapidly. Continuing to follow the model of AUI, the thirteen founding universities of UCAR achieved legally incorporated status a month after the Blue Book release. The acronym UCAR thereupon stood for the University Corporation for Atmospheric Research. Henry Houghton remained as chair of the new entity, and the represented universities provided the trustees. UCAR then had a legal status with which to enter into negotiations and various agreements with other organizations such as the member universities and the U.S. government. With the Blue Book, UCAR now had a general idea as to what form the NIAR should take.[66]

The Blue Book set forth an ambitious goal of a major center that could "mount an attack on the fundamental atmospheric problems" by providing facilities and interdisciplinary talents "beyond those that can properly be made available at individual universities." The proposed institute, it suggested, had the "possibility of preserving the natural alliance of research and education without unbalancing the university programs."[67]

To facilitate the process, the authors spoke in detailed terms about the new site for atmospheric science, suggesting that this "straw" (quotes in original) institute enable the planning and budgetary process to continue while leaving open the opportunity for change.[68] This action allowed the creators of the new institute to drift further from the original motivations that led to the calls for a new meteorological institute. As a result, weather modification appears in the Blue Book only in passing. This was an interesting and important omission given the impetus for the new center in the Orville report and the USWB weather control woes earlier in the decade.[69] But the omission may reflect the mainstream scientific attitude that weather control was an uncertain enterprise at best.[70] One important idea that the authors did emphasize concerned the desire that the institute have a strong interdisciplinary focus.[71]

Although welcoming the UCAR report, NSF leadership expressed more caution about embracing the plan that prefigured a major new meteorological enterprise in the United States. In March, UCAR submitted a proposal based on the Blue Book to NSF for $33,253,000 spread over a six-year period.[72] NSF's senior advisory National Science Board, headed by NAS's Detlev Bronk, had reservations about such a major national initiative. In their first pronouncement on the COM report and the Blue Book in May 1959, the NSB avoided the issue of a meteorological institute, rather focusing on recommendations for further support for atmospheric research at existing university programs.[73]

This National Science Board position represented perhaps retrenching by Bronk. After all, he had created COM in 1955 and enthusiastically supported the original COM report in February 1958 that called for a new institute. The board expressed uncertainty as to the exact purpose of the institute, in addition to concern about the large commitment of funds by NSF fostering the rapid growth of such an institute. By emphasizing support of individual scientists, the advisory NSB showed it did not feel comfortable yet in taking "a plunge into big science."[74]

Fortunately for UCAR and their future institute, NSF director Waterman and a deputy, Randall Robinson, "were really dedicated to the concept of NCAR." "Without their absolutely wholehearted commitment" thought Roberts, the institute "never would have happened."[75] As Waterman had started to shape U.S. biology, thereby plunging NSF into what some call "big biology," he also was then steering NSF in the late 1950s into "big meteorology" via his support of the government-funded meteorological institute.[76] This action fit well into NSF's broader goal to develop large-scale national scientific facilities.[77] Although much of this effort had an astronomical focus, Waterman's support of the atmospheric scientists and meteorologists kept advancing NSF toward assuming a major, if not dominant, role in establishing U.S. science policy over all the natural sciences—an ambition Waterman held from the beginning of his tenure as the agency's director.[78]

Despite the reluctance of the NSB to enter the world of big meteorology, the NSF did have $500,000 in its budget for fiscal year 1960 that allowed the board (and Waterman) to continue supporting work on the new institute. The NSF was hedging its bets.[79] Paul Klopsteg, previously of the NSF and then chair of COM, remained as a "special advisor" to Waterman. In a memo to Waterman, he expressed what appeared to be a growing opinion in the U.S. scientific community. Observing the rapid developments in government support of basic research activities since 1951, he concluded that "research institutes operated by universities with complete Government financing will become an accepted part of our pattern of basic research within a very few years."[80]

In a subsequent NSB meeting in August, the board authorized the use of the $500,000 allowing UCAR to hire a director and small staff to plan programs and facilities in consultation with member universities. The NSB did not really foresee, or want, a major stand-alone center. It desired an institute that would simply augment and assist existing university research in atmospheric science and meteorology.[81]

Houghton initially expressed "disappointment and shock" over the board's decision to the UCAR trustees. He came to view the NSB action not as a stoplight, but as "at least a flashing yellow."[82] About this time, in part because someone noticed that "NIAR" was rain spelled backwards, UCAR began calling the new entity a "center."[83] Soon after, NIAR became NCAR— the National Center for Atmospheric Research. Detailed planning on the new, but still poorly defined, center for atmospheric sciences began.

In his letter to Waterman about NCAR, Klopsteg commented that "the preliminary plans are tentative though impressive," but that a "competent director," when found, would review and change the plans from their "first approximation."[84] The primary task confronting UCAR trustees in late 1959 therefore concerned selecting a director to help define for the organization that existed only on paper.

The trustees understood that the new director would have great influence over the ultimate success or failure of the new center. Given the importance of NCAR to the U.S. community of atmospheric scientists and meteorologists, the trustees did not particularly care if the first director came from among their ranks. Rather, they wanted a scientist–administrator who would give NCAR the best chance for success. Their widely cast net soon fell upon the interdisciplinary Roberts and the burgeoning sun–earth network of scientific activities in Boulder.

"The Thin Ranks of the Atmospheric Sciences"

Walter Roberts had developed scientific interests beyond solar astronomy in the 1950s. He focused increasingly on the sun–weather aspect of the broader sun–earth connection. This drift from meteorology, or "accretion" of interests as he later called it, put him in contact with the community of U.S. atmospheric scientists.[85]

Roberts had created a noticeable meteorological component of HAO with the recruiting of up-and-coming atmospheric scientists such as a PhD graduate from New York University, Julius London. Roberts also convinced Bernhard Haurwitz, a pioneer in theoretical atmospheric dynamics, to migrate west. Even though the work of ISTR did not easily fit into the mainstream of atmospheric science and meteorological research of the time, it did help make Roberts and his efforts a high-profile fixture in the evolving discipline of U.S. atmospheric science.

Roberts's work on sun–weather connections had a place in meteorological thought similar to the then-current views on weather control. Neither was considered on the fringe, but atmospheric scientists expressed great uncertainty as to the outcome of the research in both areas. Both subjects were considered at least worth of study. If for no other reason, many scientists thought, such problematic investigations might result in a better understanding of basic atmospheric processes. This would be true even if such work led to research dead ends. Roberts created the privately funded ISTR precisely to investigate some of these basic sun–weather connection processes and how this knowledge might assist in routine weather forecasting.

The Institute for Solar Terrestrial Research, the High Altitude Observatory, the Department of Astro-Geophysics, and IGY activities kept Roberts busy and highly visible during 1957–58 in national and international scientific communities of the time. Even though permanently based in Boulder, and far from the eastern scientific establishment, by 1958 Roberts had established a sterling national reputation in the highest levels of U.S. science. Many viewed him as an able scientist, scientific administrator, and science fundraiser with strong interdisciplinary interests across the atmospheric and space sciences. As his close association with Sydney Chapman showed, many of these high opinions about Roberts's abilities reached abroad as well.

Attesting to Roberts's increasing stature in the atmospheric science world, UCAR and NSF appointed Roberts to two positions of significance in early 1959. First, UCAR selected Roberts as one of the first three trustees-at-large of their new corporation in April of that year. They chose him in part because he participated in the conferences that led to the creation of the Blue Book and had impressed the UCAR board.[86] Of the fifteen trustees selected at the time, Roberts stood out as one of the few who did not have a significant background in meteorology.[87] Despite this, the meteorological and atmospheric science community had come to trust him and solicited his views on how to shape this new center for interdisciplinary studies relating to atmospheric research.

Second, Alan Waterman created yet another D.C.-based advisory panel on atmospheric sciences, this one to support NSF's still relatively new and developing program in the discipline. The head of the NSF asked astronomer Roberts not only to serve on the panel, but to serve as its first chairman. Roberts agreed, and then had to resign as trustee of UCAR after only a few months because of potential conflicts of interest. He thought in this new position he might better assist both UCAR in its development of NCAR and atmospheric science in general.[88] Demonstrating the close-knit sun–earth community of which Roberts had become a prominent part, his replacement as UCAR trustee-at-large was the first chair of the NAS Committee on Meteorology, the ubiquitous Lloyd Berkner.[89]

Roberts presided over the first meeting in 1959, and the panel discussed the UCAR proposal for the new national center as one of its first actions.[90] Roberts had reached prominence in U.S. atmospheric science policy and Washington-based decision-making circles. He also had gained intimate familiarity with UCAR, many of its members, and the ongoing creation of NCAR. He was well poised for UCAR to consider him as a candidate for NCAR's first director.

UCAR desperately needed a director to advance the center, as they had received the "flashing yellow" light from NSF with the NCAR planning money.[91] When they began casting seriously about for this important position, Roberts's name appeared on a candidate list. UCAR did not put Roberts's name first, however.

James Van Allen, still flush in the glow of the 1958 discovery of his eponymously named radiation belts, headed the UCAR list of potential directors. UCAR created a nominating committee as early as April under the leadership of Horace Byers from the University of Chicago. By August 1959, Van Allen, nominated by the trustees, became UCAR's top choice for the position. He expressed interest in the job to Byers.[92]

By November, for unknown reasons, Van Allen had a change of heart. Possibly he thought accepting the directorship might require him to move from the University of Iowa where he had set up a major space science laboratory. The UCAR nominating began anew. A short list of four candidates emerged—Michael Ference, Jr. on the board of UCAR trustees, Herbert Friedman of the Naval Research Lab, the Blue Book's Thomas Malone, and Walter Roberts. Of the four, Malone alone had a formal academic background in meteorology or atmospheric science.

Neither Roberts nor Van Allen had "insider" status in meteorology. Van Allen focused his scientific interests in the upper atmosphere and beyond, and so had even less to commend him to the atmospheric scientists and meteorologists than Roberts. The fact that UCAR largely went outside the atmospheric science and meteorological community of the time to search for directors of this vitally new disciplinary organization allowed for Roberts's nomination. Why were there not more atmospheric scientists on this short list?

"Roberts's selection . . . reflected the thin ranks of the atmospheric sciences at the time" opined atmospheric scientist Robert Fleagle in a memoir years later.[93] He referred to the dearth of highly qualified candidates for the NCAR directorship from within the atmospheric sciences. Few, even those most distinguished in the field, had run anything more than an academic department. Most of the many meteorologists trained in WWII who had run organizations of any size had tended to stay in operational, not research, activities after the war. Many had left the field entirely. As noted by COM, atmospheric science faced something of a crisis in this regard—a shortage of significant research activities and scientists to undertake them. Adding to these issues was the lack of experienced administrators in the discipline to manage large research facilities. A prime motivation for creating NCAR derived from the widespread and increasing desire in the atmospheric science community to address these problems.

Additionally, as a still maturing scientific discipline, American atmospheric science had not developed many research activities or centers larger than a university department. The first numerical weather prediction centers had only just begun forming in the 1950s.[94] Because of the state of the discipline, UCAR did not have a supply of capable candidates from within for NCAR director. This left the door open to Roberts and the others that UCAR trustees deemed of sufficient ability and prestige to undertake the assignment. This dearth of scientists with the stature and experience to run the new center underscores that meteorology in the mid-1950s was still a developing discipline in many ways.

Malone and Ference indicated to UCAR that they had no interest in the job, citing various personal reasons. Friedman appeared only "moderately interested" after a "tentative approach" by UCAR.[95] Even Roberts, when contacted about the position by UCAR member A. Richard Kassander in late 1959, demurred as well.[96] Although interested, he did not think it wise for him to accept it at that time. He thought the position might take him from Boulder and his research.

Roberts and Kassander gave slightly different accounts of the process that led to Roberts's eventual acceptance of the directorship. They both agree that Roberts initially refused the position, and Kassander remembered that Roberts "indicated that I was wasting my time."[97] Their stories diverge somewhat in what happened next, but both agree on the importance of University of Colorado president Quigg Newton in convincing Roberts to head NCAR.

In Kassander's recollection, after insisting Roberts think more about the offer, it was Roberts who suggested that Kassander come to Boulder. Roberts also proposed that "we lay it all out on the table in front of Quigg Newton."[98] Roberts, on the other hand, recalled that after his initial refusal, Kassander "was lining up allies to talk me into taking the job," including Newton and NSF director Waterman.[99]

Regardless of which account was more accurate, the three men met at Newton's house in early 1960. Roberts restated his desire to remain in Boulder and with HAO. Apparently, Newton magnanimously suggested that he had no problems with Roberts leaving the university and taking HAO with him as part of NCAR. Newton believed that for Roberts "this was an opportunity to contribute to science which, despite his qualms, he could not turn down."[100]

No doubt Newton also had in mind the potential effects of bringing NCAR to Boulder and its benefits for not only the city, but his university as well. NCAR's permanent location still remained an open question. As early as December 1958 and before the Blue Book came out, Newton wrote UCAR about the possibility of locating the proposed NCAR in Boulder.[101] The shrewd politician–university president probably understood that NCAR, with Roberts as director, stood a good chance of coming to Boulder.

Newton's insistence might have convinced Roberts to accept. Kassander later thought that Roberts's "apparent reluctance" formed part of Roberts's "campaign" to generate an "eagerness of others to take up his cause." "I still don't have the foggiest idea of who really influenced whom," Kassander mused later.[102]

There were still issues to address before Roberts became director. He had developed a differing view from that of UCAR's trustees with respect to how NCAR should eventually operate as a research institution. Roberts envisioned a scaled-up version of the autonomous HAO, with individual scientists pursuing more or less their own projects.[103] In December 1959, Roberts wrote to NSF atmospheric science program manger Droessler that it seemed to him that there was a "preoccupation" with UCAR on restricting the proposed center "to problems that cannot be undertaken in universities." Of course, this "preoccupation" Roberts thought existed came from the founding members of UCAR. Rather, he argued, that the "proper goal is a strong . . . group of hard working research men" who would have the support "to allow them to fulfill their major research ambitions."[104]

Roberts's view stood in sharp contrast to the idea of NCAR as a place for research that university departments could not undertake because of scale and resources required. UCAR's original idea focused on NCAR as more of a compliment to university-based atmospheric science research. This approach for the new center avoided direct competition between NCAR and existing departments for staff and funding.[105]

Despite these divergent views on the purpose of NCAR between Roberts and UCAR's Executive Committee, the Blue Book recommendations had left maneuvering room in NCAR's final structure, goals, and location. Because of this latitude, all parties felt comfortable enough after subsequent meetings to move forward with Roberts as director.[106] Their acceptance of a director with such differing views on the very purpose of the institution demonstrates the urgency that the UCAR board felt in getting their new research center organized. They understood that the time was right and that the situation might change before they acted. It also demonstrates that they thought Roberts was worth the risk to their idea of what NCAR should become institutionally.

Roberts agreed to stay at least "long enough to get it going," and that he could leave "in a year or two" if things did not work out for him or UCAR satisfactorily.[107] UCAR had someone as their inaugural director who had proven himself as a very able, proven institution builder, science administrator, and fundraiser.[108]

Despite any lingering reluctance because of Roberts's diverging views on NCAR's purpose and structure, UCAR trustees quickly and unanimously approved Roberts as the first director of NCAR in April 1960. Roberts's enthusiastic support of NCAR, combined with a "persuasive personality," eliminated any reservations from UCAR officials about him, thought UCAR trustee Thomas Malone.[109] Given Roberts's frequent assertion that he had no plans to leave Boulder, by voting for Roberts the trustees no doubt also realized they had not just voted on a director but mostly likely on an NCAR site as well—Roberts's home city, Boulder. How Roberts, UCAR, and NSF got NCAR to Boulder follows.

CHAPTER EIGHT

NCAR and Boulder's Entry into the "Environmental Era"

The previous chapter discussed important events in Washington, D.C., that reshaped U.S. atmospheric science policy. This altered science policy opened the path to a national research center for atmospheric science. As O'Mara points out, often a focus solely on activities in Washington may skew our understanding of the role of local forces in the growth of cities of knowledge. Such a focus may lead us to neglect these immediate factors in the development of scientific sites.[1]

This chapter analyzes how regional politicians and Boulder citizens of many stripes once again aided in bringing science to Boulder in the form of the National Center for Atmospheric Research. The chapter also demonstrates how the foundation of sun–earth connection science enabled the city to thrive in the 1960s as a place for astrophysical research. All this research in the atmospheric and space sciences then served as the foundation for the subsequent burgeoning of environmental science activity in Boulder.

The creation of NCAR was one of the major institutional developments for geophysics in North America. This event was on par with the relocation of the Smithsonian Astrophysical Observatory to Harvard, the creation of the Scripps Institution of Oceanography, and the International Geophysical Year itself.[2] As a result, NSF's decision to locate NCAR in Boulder was a major coup for the city's development as a place for the production of scientific knowledge.

"You Could Find Yourself Another Director"

On June 27, 1960, Alan Waterman and Henry Houghton announced the appointment of Walter Roberts as NCAR inaugural director. In addition to announcing the $500,000 contract with UCAR to establish NCAR, an NSF press release also stated that "once established," Roberts's staff "will draw up specific recommendations concerning location and characteristics of permanent facilities required."[3] In a case reminiscent of the process by which the NBS labs got to Boulder a decade before, UCAR, Roberts, and local Boulder forces began a significant effort to get the new NCAR research facility located in Boulder.

Roberts's desire to stay in Boulder came as no surprise to the UCAR trustees. When UCAR representatives asked him during a preselection interview what he would do if NCAR did not wind up in Boulder or HAO did not eventually become part of NCAR, he replied "you could find yourself another director."[4]

What had to transpire, as was true of the NBS move a decade before, was an objective process for site selection that UCAR and NSF could claim as the basis of their decision should any objections arise. NSF, as the government funding agency, could find itself especially vulnerable to irate congressional representatives if they could not coherently defend the location selected.

The UCAR site committee set the criteria for the decision in October 1959. The criteria included categories such as accessibility, diverse weather phenomena, satisfactory living conditions, cultural amenities, scientific suitability, and a proper intellectual atmosphere.[5] Boulder, or any other potential site for that matter, had yet to surface. Four broad geographic areas resulted from their considerations. One sat in the Northeast, roughly centered in New York State. Another area was in the Southeast, centered in North Carolina. The third was a triangle centered near the University of Michigan. East of and along the Rocky Mountains was the final region under consideration.[6] The site committee then suggested that, given the general criteria, a decision should await the selection of a director.[7]

Boulder fit easily into these criteria, but so did many other locations. By the late 1950s, the city had an international reputation as a center for the generation of space science and atmospheric knowledge via sun–earth research. Perhaps Boulder's obvious fit with the site criteria and its prominence as a place to do this type of science made it easier for UCAR to pursue Roberts. Roberts's assertions that he would never leave Boulder did not really matter to Houghton and others at UCAR—they wanted him as director and Boulder could serve well enough as the new scientific center's home.

Boulder made sense, irrespective of Roberts's strong desire to have it located in his hometown, so a Roberts–Boulder package did not cause concern for UCAR's leadership.

Once Roberts settled in as director, he assisted UCAR with the site selection process. At all times he steered the decision toward the Colorado site, fulfilling the claim he made to UCAR that if selected as director he would "immediately start working as hard as I could" to attain his personal objectives for NCAR—including "Boulder and so on."[8] The UCAR site selection committee, reconvened in July 1960 after Roberts officially assumed his new duties, worked closely with Roberts to get the center located in Boulder. The question became, even well before the site selection committee gave its final report to NSF recommending Boulder, what location in Boulder worked best for the new institution.

Roberts already had a location picked out high on Table Mesa (sometimes also referred to as Table Mountain) just south and west of the city. The mesa stood not far from the National Bureau of Standards labs that Roberts aided in relocating to Boulder earlier in the decade. The UCAR site selection committee and Roberts worked closely together to produce an overarching site report justifying Boulder as NCAR's home while simultaneously driving the definitive NSF site decision to the exact spot that Roberts had previously decided upon.

Even before Roberts's selection as director, a number of localities and academic centers around the United States got wind of the new center. Many expressed a desire to host the site. As Thomas Malone commented about this time, "everyone had a small or large dream" with respect to the new NCAR facility.[9]

It came as no surprise that officials representing these locations contacted Waterman, UCAR members, and others involved in selecting the site. On June 28, only a day after the official public announcement of the revised NCAR site selection process, Senator B. Everett Jordan of North Carolina called the NSF director. The senator stated that he "wanted to be certain the UCAR would visit the Research Triangle before making a decision."[10] Connecticut-based UCAR trustee Malone found himself lobbied by a group that pressed for a site in nearby New York State.[11] Quigg Newton suggested the (non-UCAR member) University of Colorado campus as a possible host as early as late 1959. A number of UCAR member universities also contacted either NSF or UCAR directly on the matter. UCAR member and Penn State president Eric Walker, for example, expressed a strong desire as early as 1958 to host the new research center.[12]

With this high-level interest, the site decision might come under outside scrutiny. The decision could evolve into a contentious political issue for UCAR and the National Science Foundation as well. This possibility lurked in the background for Roberts, UCAR, and NSF in the second half of 1960. Although they wanted to proceed expeditiously as they all wanted Boulder as NCAR's home, they also had to proceed cautiously.

The reestablished UCAR site committee seems to have conducted its own investigations on possible locations of interest. This effort built on the previous work of the committee done before Roberts's selection. It appears that no formal call for site proposals ever emanated from the committee to interested locations.[13]

The final November 1960 site report simply stated that "the Chairman of the Board and the Chairman of the Site Committee kept complete files of reports, letters, and memoranda from a number of institutions, communities and individuals asking consideration." The report contains no list of what locations did get serious consideration or site visits (if any). It further states that as early as July 18, the UCAR Executive Committee instructed the Site Committee "to evaluate a number of sites in the Boulder area," and requested a report to the Executive Committee by October 1960.[14] An internal 1963 NSF review of the process by which the foundation helped to create the new center concluded that "no spot other than the immediate vicinity of Boulder, Colorado, was really studied."[15]

The UCAR Committee, at their October 11 meeting, recommended not only Boulder as NCAR's new home, but the exact spot on Table Mesa where Roberts wanted to build a technological center "to rival Cal Tech" well before UCAR came into being.[16] The final document also made no mention that the preference of the director was a large factor in site selection. It did mention that "the Director, who had also given careful study of the qualifications of the site, expressed wholehearted approval of the site choice."[17] Some at NSF expressed enthusiasm for the Boulder location. Randall Robinson, one of Waterman's deputies and a supporter of NCAR and Roberts, wrote the NSF director that "the situation concerning a possible site has developed rapidly, and, I feel, in a happy way."[18]

Not all thought the situation had developed in this "happy way." Houghton of UCAR and high-ranking officials of NSF maintained concerns about the site selection process. In a memo to NSF atmospheric sciences program manager Droessler on November 2, 1960, NSF Assistant Director of Administration J. E. Luton expressed the reservation that the work done by UCAR "is not adequate to support a recommendation from the Director." He thought this because "it appears . . . that equal consideration was not given to each of the four broad areas which the site selection committee initially selected for consideration." He added that three of the areas "seem to have been eliminated on the basis of the preference expressed by the Director of NCAR." Luton noted there was strong interest by congressmen in the eventual location of NCAR, including the powerful Albert Thomas of Texas. As a result, he concluded, UCAR needed to build a better case for Boulder in the event of a challenge from local and state officials.[19]

Waterman, although supporting the Boulder recommendation, agreed with Luton's analysis. A few days later he wrote to subordinates that UCAR's report needed "strengthening" as it "strongly gives the impression that their job was to justify the Boulder site, and that this was related to Roberts's stated interest."[20] Waterman knew he had to defend the choice to NSB at their mid-November meeting. He needed all the support he could gather for his argument.

Houghton already had sensed unease about the way the site selection had come about. He had started thinking that NSF did not have what it needed to defend the Boulder decision adequately even before Luton's memo. Roberts knew of his concerns.[21] When Houghton submitted the first version of the site selection report on October 28, he assured Waterman that all decisions on site selection stood squarely upon "objective" criteria, using the word four times in one paragraph alone to drive home the point. He also told the NSF director that UCAR had received "a considerable number of site proposals" that the selection committee had evaluated using the Blue Book criteria. Despite these assurances, Waterman still did not think he had enough, as evidenced by his reaction to Luton's memo on November 2.[22]

In the interim, Roberts had Tician Papachristou, an architect from Boulder, develop a location rating chart based on Houghton's somewhat expanded version of UCAR criteria. Only Boulder, Sterling (New York), and Research Triangle (North Carolina) figured in the rating. Boulder scored far ahead of the other locations by Papachristou's reckoning.[23] Roberts showed the work to Droessler and others at NSF during his visit on October 24–25, 1960. Droessler did not want Papachristou's work to appear in the UCAR report to NSF because if it were shown around "it would be dynamite"—it seemed to show that UCAR stacked the deck in Boulder's favor.[24]

If those involved had seen Papachristou's private documents, more "dynamite" might have resulted. Dated October 24, the architect wrote in his notes, "1) Make comparative list or 2) make list of criteria and fit *only* Boulder to them" (emphasis in original).[25] Either way, Boulder could only come out on top.

Waterman's continued wavering on a Boulder site produced a flurry of activity for Droessler, Roberts, and others in late October and early November. Adding to these concerns, President Walker of UCAR member Penn State—also a member of NSB—appeared "out of joint" about the selection process. Even Droessler "was a little shocked" when he found out that Penn State's site proposal "had not been officially brought up." Walker would require "considerable missionary work" before the NSB meeting in November that ruled on the UCAR recommendation. Some suggested ionospheric physicist Art Waynick at Penn State, a colleague of Roberts, as a candidate to undertake this sensitive mission to assuage Walker.[26]

On November 2, Roberts and Droessler attempted to develop a strategy on how to proceed in the weeks before NSB made its site decision. Droessler thought it "very urgent" to have Roberts come to D.C. to boost, or in his words "screw up," Waterman's courage. In addition to the report's problems, Congressman Thomas had "kicked around" some of those working on NCAR, and "they" (presumably Thomas) "are trying to slow down the whole UCAR thing." Roberts, in addition to reworking the report, needed to generate "some positive enthusiasm" at NSF.[27]

Waterman wanted, according to Droessler, "something that will give a little more discussion scientifically speaking, of the underlying weather or atmospheric problems as they relate to the site." The possibility of close scrutiny from Congress—Congressman Thomas in particular—made them all uneasy. Waterman was "nervous about sitting down with Mr. Thomas without a weather facts book." Roberts agreed to set about improving the site report.[28]

Roberts then talked to Henry Houghton of UCAR about these problems. Houghton expressed concern about inserting a climate and weather study in the report at such a late date. He thought this was "asking for trouble," probably as such a study might put other regions in the United States that met these meteorological criteria back into consideration. Roberts said he would "try to make it non-controversial."[29]

While Roberts worked these issues, Waterman had another important fact he could use in defense of the Table Mesa location that perhaps helped him "screw up" his courage when he faced NSB—Steve McNichols, Governor of Colorado, offered the site to NSF for free. In a move reminiscent of the Boulder Chamber of Commerce's grant of the land for the NBS labs a decade earlier, Colorado forces mobilized from early on in the process to come up with land. This process began immediately after Roberts's selection as NCAR director.

As early as July 12, Roberts met with long-time acquaintance Franny Reich and the Boulder Chamber of Commerce to plan a strategy for their campaign to get NCAR to Boulder. Roberts thought "fast action" could "forestall the development of strong pressures from elsewhere." The Chamber representatives heartily endorsed the effort, offering money to assist. Colorado recently had donated the Colorado Springs site for the U.S. Air Force Academy to the federal government. Those involved in bringing NCAR to Boulder agreed this course of action could work again for the new meteorological institute. Roberts agreed to work with the University of Colorado and the governor to get the state government to donate the land. Roberts agreed to arrange for the UCAR site committee to visit the city. He planned to show them possible locations and tell them "that one of these as they wish will be given to UCAR." Roberts had to have all these commitments and arrangements in place by October 11, the date of the UCAR site recommendation board meeting.[30]

Governor McNichols did not require encouragement in supporting NCAR. It was he who introduced the proposal on the need for a national center of atmospheric science at the May 1958 U.S. governors' meeting.[31] Although there is no clear evidence as to why he did this, McNichols had a very good relationship with Roberts. McNichols may have heard about the NAS Committee on Meteorology report recommending a new research center from the influential Boulder scientist. Roberts understood that the governor could serve as an important ally in his quest to get NCAR not only to Boulder, but to the Table Mesa site Roberts so desired.

"Dr. Walter Orr Roberts has briefed me," McNichols wrote to Waterman on October 7, about "the plan" of the University Corporation for Atmospheric Research to locate NCAR in Boulder. The point of the letter, he wrote, "is simply to assure you" that if NSF decided on the Boulder Table Mesa site, the "State of Colorado will undertake to provide it" either to UCAR or NSF. Waterman responded to the offer on October 22 in a very neutral tone, commenting that he appreciated the offer. Not tipping his or UCAR's hand, he added "that the Colorado locations you mentioned in your letter will be given full consideration."[32] The noncommittal tone in these letters belies that the two officials certainly knew that Table Mesa had no real competitors in Boulder, or anywhere else. For all intents, once UCAR's board picked Roberts as NCAR's first director, they also picked Boulder as NCAR's home.

Armed with a reworked site selection report and the promise of state-donated land, Waterman and his subordinates faced the sixty-eighth meeting of the NSB on November 17 and 18, 1960. The work of Roberts, Droessler, and others paid off. Concerns about the site decision that had surfaced within NSF in the preceding few weeks had faded. NSB agreed to the Table Mesa site recommendation of UCAR as the future home of the nation's first major atmospheric science research facility.

Local reaction on the Boulder side was immediate and triumphal. "We dood [*sic*] it! More confusion to the enemy!" wrote Roberts's assistant, Mary Andrews, in a handwritten memo to Philip Thomson, NCAR's deputy director, on November 22. "Onward and upward to the next battle!" she proclaimed in the note, further expressing widespread local sentiments on the NSB decision.[33] "The enemy" to which Andrews referred in her memo remains uncertain. The next battle presumably concerned the drive to amend the city's so-called "Blue Line" so that public water service could get to the NCAR site.

Waterman and Roberts had to engage in preemptive political damage control concerning fallout from the location decision. These conciliatory actions worked, for there were no challenges to the Boulder decision coming from representatives of competing locations.[34]

The "missionary work" Droessler thought necessary for Penn State's president Walker also proved effective, for no disturbance came from that quarter either. Art Waynick and possibly UCAR members on Penn State's meteorology faculty apparently convinced the recalcitrant president and NSB member on the possible long-term benefits NCAR might have on his university, irrespective of where the new center was located.

The next "battle" focused not on getting the site from the state, but rather on simply getting water to Table Mesa so people could work there. Once again reminiscent of the NBS move earlier in the decade, the people of Boulder had a voice in the question of bringing science to their city.

Boulder Votes for Science

"Cities of knowledge," wrote O'Mara, "are the product of local action," and the Boulder story bears out this observation.[35] The local action this time existed in a more complex local political environment than when Boulderites banded together a decade earlier to get the NBS lab relocated to the city. Local thinking about the best way for the city to grow had changed in Boulder since the NBS move. Enthusiastic pro-growth ideas of the late 1940s and early 1950s had given way to concerns about the deleterious effects of unrestrained growth on the picturesque city. Roberts's dream for the new NCAR building ran straight into one of the barriers the city had recently erected to control growth—the "Blue Line." In an ironic twist, Roberts's wife,

Janet, had helped to create this final barrier to getting a major piece of big science into Boulder.

The Blue Line began as an informal attempt in 1959 by a number of Boulder's citizens to preserve the city's scenic beauty and perceived quality of life. Robert McKelvey, a math professor at the University of Colorado, contacted his friend and professor of physics, Albert Bartlett, after hearing of a planned development of a resort hotel on one of the mesas that adjoins Boulder immediately to its west— part of the region known locally as the Flatirons. "We have got to do something about it," a determined McKelvey told Bartlett.[36]

Janet Roberts, who had become an appointed member of the Boulder Planning Commission in 1956 and then an elected member of the city council in 1959, soon decided to join those who wanted "to do something about it." She felt persuaded that development on the mesa was potentially "tragic." She joined the growing local movement to restrict growth, especially growth up the foothills to the west that might mar the scenic beauty of the location.[37]

Boulder in the 1950s, according to a council member at this time, Frank Havice, "was controlled by pro-growth factions. Growth was in the same category as motherhood."[38] The anti–rapid growth elements of the population quickly had gained strength to the point where they could bring significant voter pressure on important decisions for the city's future development.

A petition instigated by these controlled-growth advocates forced the city council to call a special election to consider a measure that would alter the city's charter to disallow any services to the Flatirons. This citizen-generated measure competed with the council's measure for a water bond that provided for water services up to the mesa areas. The limited growth proposal set a limit at 5,750 feet in elevation—above this, Boulder would not provide water.

In the special election held on July 21, 1959, the voters of Boulder defeated the water bond and approved the resolution that amended the city charter to forbid the extension of water service. The new 5,750-foot elevation limit entered local parlance as the "Blue Line."[39]

To continue the momentum generated by this victory, just about the time Janet Roberts assumed her position on the city council, Bartlett and others created the People's League for Action Now. The group acquired the shorthand PLAN-Boulder and became something of a permanently standing local watch group on unrestrained growth.[40]

A little less than a year after its passage, the Blue Line limit to growth and the PLAN-Boulder vision for the city clashed with Roberts's vision for the growth of science in Boulder on a mesa and above the Blue Line. Yet, subsequent events showed how the idea of Boulder as a world center for atmospheric and space science via the original sun–earth connection work had become part of the city's self-identity. This idea of Boulder as a special place for science existed even among the PLAN-Boulder activists, including Janet Roberts. This identity coupled with an odd circumstance—by allowing an exception for the building of NCAR the city actually helped to preserve the area's beauty in the longer term. This confluence made the choice for the citizens of Boulder to amend the Blue Line only for NCAR straightforward.

Other municipal sources expressed a desire to provide water to the foothills at this time. If successful, these actions could have opened up areas behind Boulder for possible development because the water coming from non-Boulder sources negated the purpose of the Boulder's self-imposed Blue Line.[41] Robert Turner, city manger of Boulder, expressed concerns on November 10 at a meeting of NCAR representatives, university officials, and others that "Denver is definitely in the water selling business." He feared that "the developers will bring in Denver water and quite possibly incorporate as a separate town at the edge of Boulder."[42] Boulder ran the strong risk of having other municipalities control water to the area surrounding Boulder if the city held firmly to the Blue Line with respect to the NCAR site.

Roberts understood early on that his handpicked spot on Table Mountain presented problems because of the Blue Line restriction. Even before UCAR approved the site in October, he discussed the situation with one of McNichols's aides. Roberts should do some "public relations," suggested the aide, adding, "this would pave the way to getting the various conservation minded groups behind some kind of a move to supply satisfactory water facilities to the site."[43]

This placed the Boulder city government in a bind. Although it could not guarantee services to the planned NCAR site because of the Blue Line amendment, the city also did not want the site, once obtained by the state, to have its water provided by Denver or any other municipalities other than Boulder.

This bind also brought an opportunity. By this time, the Boulder city council, except for Janet Roberts, who recused herself from any NCAR-related business before the group, decided to have citizens vote on amending the Blue Line for the purpose of a one-time exception for NCAR. Allowing NCAR to build on the site was a small concession if this blocked others from encroaching on the city.

City manager Turner thought that "even the PLAN-Boulder people and other dedicated conservations" would have to support the Blue Line exception. The exception represented "the only practical means of forestalling an explosive development" of the mesa area and other regions adjacent to

Boulder. Others advocated for a natural preserve status around the site as part of the NCAR development.[44]

Joseph Rush, a long time Roberts associate at HAO NCAR, added another factor that continued to resonate in the amendment process: "Every scientific establishment that locates in Boulder automatically increases the attraction of the area for other such institutional enterprises," putting out an idea that he knew would have traction for many of Boulder's populace. The combination of the ideas that NCAR helped to control growth in the long term and that NCAR also attracted even more scientific activity to Boulder proved to be a potent mix.[45] Rush knew how the scientific community in Boulder had grown and how it might further grow. It seems that Rush instinctively understood the concept of a Marshallian district.

A "Citizen's Committee for the NCAR Blue Line Amendment" formed. Albert Bartlett and Frank Havice, who played prominent roles in getting the Blue Line established, were leading members. The NSF in the interim promised to set aside some of the land for a natural reserve, adding to the amendment's attraction to conservationists. Another notable group member was Frederick W. Brown, director of the NBS Boulder lab and Alan Shapley's superior.

The flyer they produced looked very much like the "prosperity insurance" flyer of a decade earlier. The entire effort was reminiscent of the previous Chamber of Commerce campaign to get the NBS labs to the city.[46] This flyer formed part of the public relations blitz that also included radio spots. Roberts and others gave numerous talks highlighting the potential benefits of NCAR's presence in Boulder, including increased employment. The Chamber may have coordinated and funded the citizen's group effort, although evidence for this is indirect and based mostly on the similarity of this effort with the NBS lab land funding campaign in 1950.[47]

Directly addressing the controlled-growth advocates, the flyer spoke of the amendment as "preserving a large portion of recreational and biologically important area below the Flatirons," and that "no other realistic prospects for preservation of this area is [sic] in sight." In a broader appeal, the flyer also spoke to the zeitgeist of the city, stating that "further identifying Boulder with education and scientific inquiry" would have many benefits. These included the attraction of "scientific and technical industry" that did not have "the objectionable features of heavy industries."[48]

A vote for the amendment was a vote for the best of all worlds—preservation of natural resources, but with the continued growth of science and education in the city. Fortunately, the NSF agreed with the idea that part of the site should remain a natural preserve in perpetuity.[49]

Few expressed a dissenting voice. Those who did found themselves in a small minority, for most Boulder citizens did not mind having NCAR on the mesa—many embraced the idea of bringing more scientific activities to the area. Those few who protested expressed concerns about the possible eyesore the NCAR building might present and went on to propose other sites in the city. One letter writer commented that "Boulder should not sell its scenic soul for the sake on any industrial development" near the Flatirons.[50] None of the few dissenters voiced any concern about science coming to Boulder —only the mesa location above the town appeared an as issue. Roberts anticipated some of these contrary views. He went out of his way to pledge to the voters on the citizen's group flyer and in his many presentations in the run-up to the election that "that preservation of the natural beauty of the site will be of the very highest importance as we design and build the center."[51]

Once again, as with the NBS fund drive in 1950, the campaign worked. On January 31, the voters cast their vote for bringing science to Boulder in the form of NCAR on Table Mountain. A majority of the voting population, over 7000 voters, turned out for this special election. The Blue Line exception for Roberts's NCAR passed by an almost four-to-one margin.[52]

The joint appeal to conservationists and to what had become a dominant self-image of the city in the 1950s—a place for sun–earth science and education—had resonated with much of the populace. Conservationists, many of whom knew Walter and Janet Roberts, perhaps were attracted by the environmental appeal of an atmospheric research center. Other individuals advocated unrestrained city growth. But whatever the motivations for voting for the amendment, the first major atmospheric science center in the nation, NCAR, came to Boulder. Moreover, it arose upon the spot where Walter Roberts long dreamt of establishing a world-class scientific research center.

The dedication of the new NCAR building did not occur until 1967, but even more science came to Boulder before then in the early and mid-1960s. In this next phase of Boulder's growth as a city of knowledge, unlike the decade and a half before the 1960s, much of this new science came without Roberts's participation. He had played a vital role in establishing the firm foundation—a Marshallian district—for science in the city upon which others built in subsequent years and decades

A Lab on the Hill

With the new NCAR and Roberts housed in temporary quarters, both NCAR and its new building began to take shape in the early 1960s. The state turned over the site to NSF in late 1961 after the state senate appropriated $250,000 to buy the land in March of that year. Colorado governor McNichols even admitted at this time that he had fudged a bit when he promised the land to NSF. Without legislature approval, the action he took was most likely "illegal."[53] Most everyone seemed pleased with Colorado getting this scientific jewel of an organization. NCAR made not just Boulder a higher "peak" of American science, but the state as well. Referring to the governor's work and reflecting a widespread view in the area, the *Rocky Mountain News* proclaimed "Steve Okays $250,000 Bill for Space-Age Science Site."[54]

Roberts, keeping his promise to Boulder citizens to preserve the NCAR location's intrinsic beauty, obtained the services of the then up-and-coming architect I. M. Pei to create a structure for NCAR that harmonized with the surrounding locale on the mesa.[55] Roberts selected Pei for his established reputation and by recommendation of UCAR and MIT's Henry Houghton— Pei designed MIT's new earth science building for Houghton. The architect also filled Roberts's criteria that the he give "part of soul" to the task of designing the NCAR building.[56] When Roberts and Pei visited the mesa site for the first time together, Roberts told the architect to make the building "inconspicuous," and added, "please save all the wild flowers."[57]

The new NCAR building, eventually dedicated in May 1967, sat upon the mesa at an altitude of about 7500 above sea level and seemingly hovered almost 1500 feet over the city itself. Leslie wrote about the design of the building reflecting Roberts's guiding philosophy of a decentralized science, a "village" consisting of a community of "self-directed" researchers.[58] He wanted a structure that stimulated an "air of ferment and intellectual cohesion that did not depend on any formal measure" for coherence.[59] Here individual researchers could work alone or in small groups, constantly exchanging ideas as they interacted throughout the building spaces in the course of a day. Pei's building, reflecting Roberts's ambitions, resulted in a complex of interconnected towers. The structure subsequently evoked to some a castle or medieval monastery.

Roberts's ideas on how NCAR conducted science, made concrete by the building, were still in stark contrast to the originally stated purposes of NCAR in the Blue Book that the facility should augment, not replace, university-based research. He was always candid with UCAR about his views on how best to do science and how he saw NCAR in many ways as a stand-alone institute. It was for Roberts not merely an auxiliary of academic meteorology. This difference generated critical comment during Roberts's tenure, and, as one participant wrote, "Walt's vision" constituted "a continuing cost."[60]

Eventually, the tension generated with the UCAR board on the best way to manage NCAR science activities led to Roberts's leaving UCAR in the early 1970s. Irrespective of the difference between Roberts and atmospheric scientists who disagreed with his approach to NCAR's function, the institute by the late 1960s had assumed a leading role in the world of atmospheric science research. Roberts, after leaving UCAR, participated in many humanistic and philanthropic endeavors. Among these included the Aspen Institute located in the mountains not far from his original Colorado home at Climax.

Perhaps the actual physical elevation of the NCAR site also reflected something about Roberts's thinking. The modernistic structure's location represented the rise of Boulder by the mid-1960s as a world-renowned city of atmospheric and space science knowledge production.

No evidence exists that Roberts selected the site with this symbolic sense consciously in mind. It is hard to escape the significance of his strong desire to have a laboratory on a prominent high point, situated against a spectacular scenic backdrop. NCAR visitors travel an ascending road to the Pei building. The selection of the site may reflect an unconscious desire of this idealistic northeasterner. Roberts grew up just a few miles from Plymouth and Boston and perhaps longed for a scientific counterpart to Winthrop's "city on a hill"—the lab on a hill as a "light unto the world." Roberts even referred to the site as "sacred" and a place the "the Ancient Greeks would have envied."[61] Pei's architecture, based on Roberts's oft-stated desires, reflected an almost monastic view of scientific research.[62]

NCAR's dedication ceremony on the mesa in many ways symbolically flagged the end of the growth of Boulder's foundation era as an internationally prominent center for scientific production—both NCAR and science in the city had reached a pinnacle of sorts. From a modest beginning as a science research center, the city had developed beyond Roberts's, Menzel's, or anyone else's imagination in 1945 into a modern city of scientific knowledge production. Upon this foundation of a scientific Marshallian district, the city soon built its position as a world center for not just sun–earth science, but also environmental research, many branches of physical sciences, and other high-technology-related research.

Boulder Science Thrives beyond the Sun–Earth Connection

Boulder's development as a city of scientific knowledge production did not stop with the creation of NCAR and its move into the city in 1960. The city continued to grow in major ways as a site for U.S. science and technology, even as Roberts's direct role waned as scientific activities in the city diversified away from his immediate interests. He focused on getting NCAR running and scientifically productive while Boulder's space and atmospheric science research expanded and diversified. Eventually, Roberts's influence as a vital force in Boulder science ended as he left NCAR and UCAR in the mid-1970s to become increasingly involved with the nearby Aspen Institute and other entities directly related to his social and cultural interests.

As many scientific centers arose in Boulder in the 1960s, the story of two major centers of scientific research in Boulder—the Joint Institute for Laboratory Astrophysics (JILA) at the University of Colorado and the Central Radio Propagation Lab—illustrate how space and atmospheric science developed in Boulder in the early 1960s in this era of government funding largess and the burgeoning of U.S. scientific institutions. So much happened in Boulder that Roberts could not have played a role in everything, even if it was his desire. The foundation provided by his sun–earth connection work provided a robust platform upon which more scientific activities built. With its science foundation so heavily constructed on sun–earth interdisciplinary work, Boulder sat perfectly situated to become a leader in the Humboldtian approach to earth science. This type of science stressed holistic, rather than reductionist, relationships in the investigation of the natural world.[63] The city as such was well on its way to establishing a reputation as a center for all types of knowledge production, not just those related to the sun–earth connection.

The creation of JILA on the University of Colorado campus illustrates the process by which the city's scientific activities grew once Boulder established itself as a major location for the generation of scientific knowledge. The sun–earth science that started with HAO in 1946 blossomed in the early and mid-1960s with many new and different entities.

JILA, like other Boulder sun–earth connection research organizations, represented an amalgam of interests between the federal and local entities. Its creation showed the creativity and diversity of institutional arrangements that often went into the organizational formation of U.S. science in this period. In JILA's case, the federal National Bureau of Standards and the University of Colorado formed the partnership.

In the early twentieth century, George Ellery Hale, creator of the Mount Wilson and Palomar Observatories, thought astronomy needed a place where scientists could do "laboratory astrophysics."[64] Lewis M. Branscomb, a senior atomic physicist with NBS in Washington, D.C., and solar astrophysicist Richard Thomas, an HAO alumni and close colleague of Roberts, came up with their idea for a similar research center at a meeting of the International Astronomical Union in 1958. Like Hale much earlier, they wanted to create a major laboratory that applied atomic physics research to current problems in astrophysics.[65] Their concept had NBS sponsoring the lab, but to have a facility located on and run jointly by a major U.S. university. Branscomb and Thomas also wanted their planned institute located far away from the distractions of Washington, D.C.

After Branscomb got NBS director Alan Astin's approval in 1961, legal issues arose concerning the proposed shared federal–university ownership and management of the lab facilities. Many D.C.-based NBS officials and lawyers did not understand how government and academic organizations could mix into some sort of joined institution.[66] Overcoming this institutional barrier proved an easy task. Showing the flexibility of U.S. science policy of the time, the Department of Commerce General Council soon told Branscomb that together they simply would "invent a new mechanism for governance" of the proposed research center.[67]

With that hurdle cleared, Branscomb began a nationwide search for potential locations. In addition to the University of Colorado, he investigated other major universities with top-flight astronomy and astrophysics programs. Included among these other potential locations were the University of Arizona, the University of California at Berkeley, and Cal Tech.[68]

Boulder and its university rose to the top of Branscomb's list. "Scientific Siberia" no longer fit as an adequate description for Boulder. Branscomb recounted that "in the early 1960s, Boulder was the most exciting place for institutional innovation in science in America," so it made sense to locate the new lab there.[69] With NCAR, HAO, and NBS, the city already had significant centers for space and atmospheric science that directly related to JILA's potential research areas. Roberts's work in conjunction with the university to create the astrogeophysics department and in getting the Sommers–Bausch Observatory built ensured that the university maintained a strong interest in astrophysics. These previous university efforts, embedded in the broader context of the growing "AstroBoulder," established a reputation for the University of Colorado as a place for cutting-edge astrophysics. As such, the school was an essential element of Boulder's growing Marshallian district for science.

University of Colorado president Quigg Newton still rode high on the tide of increased science funding. Seeing yet another opportunity to bring a major science activity to his school and the city, he wanted JILA on his campus to increase the university's growing stature as a major U.S. research center. Officials at other potential sites did not show the interest that Newton did in JILA.[70] Branscomb recalled wryly that the university president "had his eyes on the stars too, if not the Colorado statehouse, as well." Newton's close association with Roberts in the late 1950s attuned him to opportunities that arose to advance space and atmospheric science at his university. The accommodating university president worked with the regents and his finance officer to come up with ways, including loans and funded fellowships, to finance the JILA building and JILA's research.[71] The state of Colorado once again contributed a site for Boulder science, an unused National Guard armory in the city.[72]

The NBS–University of Colorado lab began operations and established itself as a major center for astrophysical research in the United States and the world. JILA's development benefited from research funds offered from the new NASA and the appointment of former NBS director Edward Condon to its staff.[73] Condon in a way finally received his "summer capital" years after he "schemed" to get the NBS lab to Boulder. Roberts "hailed" JILA's creation, also commenting on the "powerful and growing" research agenda of the university's physics department.[74]

This account of JILA's creation demonstrates that NCAR's Joseph Rush made a prescient comment when addressing the importance of NCAR to Boulder—every scientific organization that came to Boulder increased the desirability of Boulder as a center for scientific research. The scientific community that started with sun–earth connection work grew and diversified as more organizations dedicated to scientific research came into the area. They came because of Boulder's increasing reputation as a desirable location to do all types of scientific research, creating a Marshallian district for science in the process.[75] Boulder's Marshallian district then grew beyond its sun–earth science base to encompass diverse kinds of research on the natural environment by the late 1960s.

Boulder as a World Center for Space Environmental Research

Following closely on the heels of *Sputnik*, U.S. and Soviet space activities proliferated in the late 1950s and early 1960s. The need to understand the near-earth space environment rapidly rose in importance as a result—not only the military, but also many other sectors of society by then had an interest in the sun–earth connection. In addition to NASA, with its human spaceflight agenda, private communications companies, electrical power generating firms, and others began understanding that solar disturbances in the space environment might affect their activities. The exact nature of these interactions was still murky, for the IGY had only begun to provide answers to the increasingly detailed questions on sun–earth interactions. As with IGY, Boulder found itself well situated to contribute to, and benefit from, the societal need to generate new knowledge about the earth's geophysical environment.

Many governmental scientific entities understood the leverage that space environment prediction research and operations might bring to their organizations in the new space age. CRPL scientists therefore faced a challenge to their continued existence in Boulder—the U.S. Weather Bureau made a serious effort to assume the leadership in space environmental prediction. Alan Shapley, as his colleague Roberts did fifteen years before, had to directly intervene to help keep an important aspect of science knowledge production in Boulder.

Shapley, because of his work as Kaplan's deputy of the U.S. IGY National Committee, developed into a scientific administrator of repute by the early 1960s. Kaplan wrote to Secretary of Commerce Sinclair Weeks in May 1958 that the executive committee of the USNC thought Shapley's "devotion," scientific acumen, and political skill greatly increased the "scientific prestige" of not only the United States, but of the Department of Commerce as well.[76] Despite this acclaim, by the early 1960s Shapley still only held a middle-management position at CRPL. Yet, in part because of his IGY work and his participation in getting CRPL to Boulder, he knew Washington politics well despite his relatively modest official position as a division chief.

The idea arose in USWB during the fall of 1963 that it should enter into space forecasting and geophysical data dissemination. The director of the National Weather Satellite Center, S. Fred Singer, or L. H. Clark, a special assistant to the USWB director for "extraterrestrial weather" (ETW), possibly proposed the idea. By December, the discussion advanced far enough in the Department of Commerce that Singer drafted a proposal for what he called an interim organization for ETW activities. In his plan, functions were divided between USWB in Washington, D.C., and Boulder.[77]

By early January the proposal generated, according to R. W. Knecht, a division chief and Shapley's superior, a "full scale flap" between the U.S. Weather Bureau and NBS. Writing to Shapley, who at the time was in England doing graduate studies, Knecht thought that the USWB proposal did not give CRPL its "proper role" in space forecasting. CRPL's management wished to retain their status as the lead in space environment forecasting.

They wanted Shapley's ideas as one who knew Washington science politics even better than they did.[78] A serious concern developed about CRPL's dominance in space weather prediction and research, and possibly even of its long-term continued existence in Boulder.

Shapley responded that "I am with you 100% on this." But, he added as he wrote to his superior in the lab, "if I may be immodest," they needed a "policy savy [sic] person like you (or perhaps me) to carry the ball" for CRPL. He further advised that any structure developed in the short term have the qualifier "interim" attached so as to give them some maneuvering room for future adjustments.[79]

On January 11, Gordon Little, the chief of CRPL, wrote to Shapley from Washington, D.C., after meeting with Robert White, the head of the USWB, and Singer and Clark. Little reported that White most likely would suggest to Herbert Hollomon, Assistant Secretary of Commerce for Science and Technology, that the Department of Commerce should take the lead in space weather forecasting for the U.S. government. Given this, Little thought, Hollomon leaned strongly toward making CRPL (and Boulder) the lead for U.S. space environment forecasting. Showing his esteem for Shapley, Little also asked if Shapley might return to the United States to help in any subsequent negotiations on this contentious issue.[80]

Little was correct, and Shapley had no need to interrupt his studies. Almost as if to give up the fight for any role in space forecasting, White wrote to Little that Hollomon had selected CRPL as the lead agency for space environment forecasting and data collection. The Weather Bureau's role, according to White, reduced to assisting CRPL to "whatever extent they can contribute."[81] Hollomon then officially notified NBS director Astin of his decision to make CRPL primarily responsible for the department's work in "extra terrestrial weather conditions" because of CRPL's "background experience and current activities" in the subject.[82] As if to highlight the new importance of the Boulder-based facility, Hollomon soon designated Knecht and Shapley as panel members of the new U.S. government's Interdepartmental Committee for Atmospheric Sciences Panel on Space Environmental Forecasting.[83]

The infrastructure and reputation of Boulder as a science center, combined with the influence of local scientists, again helped to keep a significant component of space age (and environmental) science in the city. Boulder became a national and international leader in the rapidly developing field of space environment forecasting and monitoring—a scientific endeavor at the heart of the sun–earth connection.

Boulder and the Creation of ESSA

There were many ripple effects caused by the designation of the Department of Commerce's lead agency for space monitoring and forecasting. One of them ensured that Boulder became a major component of the creation of the Environmental Science Services Administration (ESSA) in 1965. Some scholars place ESSA's creation as another major milestone in the development of U.S. geophysical and environmental science.[84] As such, ESSA's institutional structure in Boulder further established the city as a place of scientific knowledge production. It also enhanced the city's reputation as a place for the growing environmental movement. Boulder's history as a place for sun–earth connection knowledge production poised it to benefit from a new national (and nonmilitary specific) emphasis on the environment and the resulting new government funding patterns for science. One scholar referred to the post-1960s as an "environment era," and Boulder became one of the scientific hot spots of this time.[85] Science in Boulder during the 1960s fit well into the developing awareness of the potential fragility of the natural environment and its changing nature over time.[86] In many ways, the Humboldtian approach to studying the world was Boulder science by the mid-1960s.

The Johnson administration, responding to a growing environmental consciousness in the nation, wanted a "single national focus" in the efforts to "understand and predict" the state of the planet's oceans, lower and upper atmosphere, and solid earth environment.[87] President Johnson, according to science advisor Herbert F. York, wanted American science to do something for "grandma" and not just the government or elite academic institutions.[88] His administration's creation of ESSA resulted in part from his desire for a U.S. scientific enterprise that benefited all Americans. In this case, the useful enterprise related to studying the earth's natural environment—an area of study that potentially affected all citizens.

ESSA, formed under the Department of Commerce on July 13, 1965, had major organizations transferred to it from other agencies in the Department—the Central Radio Propagation Lab, the Weather Bureau, and the Coast and Geodetic Surveys. A component of the ESSA was the newly created Institutes for Environmental Research (IES). A renamed CRPL became the Institute for Telecommunication Sciences and Aeronomy under IES. IES's first director, George S. Benton, did not hail from Boulder but was a meteorologist at Johns Hopkins University. As a member of UCAR's executive committee, he had close connections to Boulder's scientific community, including Roberts. He therefore had familiarity with scientific activities in the city.[89] As the new organization rapidly fell into place, ESSA director Robert White—former head of the USWB—pointed out to ESSA personnel that "the entire globe is our workshop."[90]

The most important aspect of ESSA's creation for Boulder did not rest in simply the reorganization and name change of CRPL. In recognition of Boulder's status as a place for this kind of scientific knowledge production relating to the earth's environment, the Department of Commerce picked Boulder as the location for the new IES headquarters. Other ESSA research activities came to Boulder as well. CRPL's Gordon Little reported that the "stated intention" of ESSA management in effect placed "all *new* research activities" (emphasis in original) in the city.[91] As a result of this expanded ESSA scientific effort in Boulder, work began on a new $600,000 research facility.

Little had justification in writing to all CRPL employees a few days before their organization transferred to ESSA that "the future looks bright for us within ESSA."[92] He could have added that the future looked bright for Boulder also. ESSA's Boulder locations were merely the vanguard of many environment-related research and data center activities that came to the city as national funding priorities shifted to broader environmental concerns. Boulder, beginning as a Marshallian district for sun–earth science, had developed by the late 1960s and early 1970s into a Marshallian district for environmental research—the new "environmental era" for the nation heralded a new era for Boulder science.

CHAPTER NINE

Conclusion

The Creation of a Modern City of Scientific Knowledge Production

From the mid-1940s to the mid-1960s, sun–earth science became synonymous with Boulder science. Some scholars discuss the rise of "science regions"—defined as the "geographical agglomeration" of facilities for the production of knowledge. These regions are not unlike Lécuyer's Marshallian districts where skilled workers assembled to generate specialized knowledge.[1] Sun–earth science in Boulder was one such regional science. This small city in the foothills of the Rocky Mountains assumed a unique identity as a national center for scientific knowledge production about the complex physical interactions that connect sun and earth.

At the fortieth anniversary symposium of HAO in 1980, Lewis Branscomb, creator of JILA in Boulder and a former director of the NBS, asked how Boulder developed as a scientific center.[2] The answer to this question is not easily rendered. Astronomer Walter Roberts played a pivotal role in this development. Many other factors contributed to Boulder's rise as Branscomb's "American Akademe-Gorodok" as well.

What makes the Boulder story interesting rests in the unexpected nature of the city's road to the top levels of the U.S. scientific research establishment. The city possessed few of the natural endowment of factors, such as a local scientific labor force, a research university, and a hub airport that might have served as a predictor for the growth of the city as a significant site of scientific knowledge production or high-tech center.[3] Unlike the growth of science regions such as Stanford–Palo Alto and major cities like New York and Washington, D.C., little existed in the 1940s to recommend Boulder as a potential center for international science. It was a "scientific Siberia," to use Alan Shapley's words in describing how the eastern scientific establishment viewed the city.

Many individuals in Boulder (and Colorado) exploited the various opportunities afforded by the changing nature of U.S. science polices and associated funding in this Cold War period—in the process, they created a "Marshallian district" for sun–earth science.[4] The list of those involved includes Roberts, scientists such as Alan Shapley and Edward Condon, educators, businessmen, Boulder residents, politicians, and government officials. There were little, if any, long-term plans or specific visions in Boulder's development as a city of science; it changed spontaneously over time. Early Cold War circumstances produced many funding opportunities that Roberts and others in Boulder exploited to their (and their city's) permanent advantage. Certainly, regional and local contexts are essential to a proper understanding of the development of the American West as a site for science.[5] Additionally, the Boulder story shows clearly that science regions are constructed by way of "the tangled circuits of social relations" that have affected what and how science was done in a particular region or area.[6]

In explaining the rise of science regions, some argued that a powerful research university provided the basis for modern cities of knowledge in this early Cold War period.[7] Yet, an important aspect of what makes the Boulder story instructive is precisely the absence of a major research university there at the beginning of the Cold War era. The University of Colorado, at best a middling state university with a modest research component, co-developed with Boulder's rise to national prominence. The story of Boulder's development belies the idea that cities of knowledge, such as Palo Alto, Cambridge, and so on, necessarily needed preexisting powerful, large research universities.[8] In Boulder, the science came first, and large-scale institution building followed.

Rather than serving as a generator of scientific activity, the University of Colorado grew alongside diverse sun–earth connection science activities in the city. Much of its subsequent success as a major research university related directly to university officials who willingly and eagerly rode the rising tides of funding and the fame of the city as a center for sun–earth knowledge production from the late 1940s. The sun–earth science Roberts brought to the city, not to the existing university, was the true catalyst in the Boulder story.

This research suggests that the mere existence of a university (irrespective of its power) provided a crucial factor in the development of Boulder into "AstroBoulder." The existing middling university was "large enough"—despite its then modest status—to provide the requisite intellectual environment to attract additional elements of the sun–earth connection network to the city. The school's mere presence served to lure Roberts down from the mountains to form the Harvard–University of Colorado HAO in the first place.

Similarly, the presence of the university—starting with just a modest physics department and only a Master of Science program in engineering—provided sufficient rationale for the NBS committee to select Boulder over the Stanford and University of Virginia regions for the Central Radio Propagation Lab relocation.[9] This again contradicts assertions that existing major research resources and a university's significant "political clout" formed the backbone of the "Cold War University."[10] The Boulder experience shows that this observation is incomplete. By focusing on existing major universities such as MIT and Stanford, scholars have often missed the creation and emergence of entirely new centers of knowledge production in this period.

The driver of Boulder's development was a compelling scientific question—the nature of the sun–earth connection. This science established the basis of Boulder's Marshallian district for science in the city. The growth of sun–earth science in the city, fostered locally by scientists, city boosters, and a series of ambitious university presidents, stimulated Boulder's growth as a science center.

The university's presidents realized that Roberts's work on sun–earth issues could do more than just form the basis of an important scientific community in Boulder. They also thought that their university could play an increasingly important part in astrogeophysics as well. These presidents realized that the development of Boulder sun–earth science could effectively contribute to their goal to build their university into a modern research center. The close relation between Roberts and each of these men does not come as a surprise, for there was a strong sense of perceived mutual benefit to all involved.[11]

Walter Roberts emerged as the most important single individual in Boulder's development in the period under study—he brought sun–earth science to Boulder in the postwar years. He subsequently transformed from a researcher into a major entrepreneur of this science in the city. His desire to remain in Boulder permanently in the late 1940s forms a crucial element in explaining Boulder's development. A few easily imagined counterfactuals based on this research clearly demonstrate this.

If Roberts did not promptly resist Harlow Shapley's idea in mid-1947 of combining HAO and Sacramento Peak activities at the latter site, it remains highly unlikely that Boulder would have evolved as it did. The nascent sun–earth connection community would have collapsed before it established itself in Boulder's environs. Without the network, what might have brought other such activities to Boulder? Boulder also would not have interested UCAR as a possible location for NCAR without the presence of Roberts and his sun–weather connection activities. NCAR found a home in Boulder when Roberts agreed to serve as its inaugural director after Van Allen declined UCAR's offer.[12] Roberts therefore exemplified Boulder science for a generation—his presence was essential to developments in Boulder.

The science activity in the city grew not only during Roberts's active career; it continues to grow to this day. These initial activities grew horizontally, not vertically, where spatial concentration encourages frequent interaction among all groups and individuals involved. The broad suite of research relating to sun–earth science stimulated this horizontal growth.[13] Roberts, not an empire-builder in any traditional sense, represented a strong break from the "imperial" laboratory director tradition dominant in nineteenth- and early twentieth-century physics, best illustrated by Harlow Shapley's rule of the Harvard College Observatory.[14] As important as Roberts was at the beginning of Boulder's development, the horizontal integration of science there ensured continued growth after he passed from the scene.

The combined story of Boulder and Roberts represents a synergistic and unique confluence of person and place to a scientific idea. This confluence, embedded in the broader issues of Cold War sponsorship patterns and disciplinary issues in the space and atmospheric sciences, enabled the growth of a prominent site for U.S. science. In effect, a Marshallian district for science developed, a district that made sun–earth science Boulder's regional *raison d'être*.[15]

Funding Patterns, Roberts, and Boulder

Roberts's scientific activities occurred over a crucial period in the history of U.S. science. Consider the era spanning from the late 1930s to the early 1960s as having four major phases of research funding. Roberts's career, tied as it was to events in Boulder, illustrates these rapid and significant changes in U.S. science funding patterns of the mid-twentieth century.

In the first phase, Roberts's graduate school years at Harvard and his time at Climax before WWII began, the primary funding came from donations and grants from private sources—an era of what one might call "private largess" for U.S. astronomy, and solar physics in particular.[16] A most unlikely donor for U.S. solar physics, the Climax Molybdenum Company, made the early solar coronagraph work possible. The company also helped to establish both Roberts's roots and space science in Colorado and near Boulder.

Limited funds for solar research came from Henry Wallace and the U.S. Department of Commerce. These funds represented a set of unusual, Depression-era circumstances and Menzel's persuasive abilities, rather than any broad U.S. government policy on science funding. Even to the pro-science and New Deal liberal Wallace, Menzel needed to argue solar physics' supposed practical benefits to the U.S. Weather Bureau. The nation was not yet comfortable with the idea of government support of basic science outside of a national emergency.

This first phase from the late 1930s to the entrance of the United States into WWII exposed Roberts to the elitist ideology that only scientists could (and therefore should) determine the direction of their research. Private funding ensured maximum latitude for the scientist in pursuing his scientific objectives, for this non-program-specific funding presumably had less political and bureaucratic interference.[17] This time of maturation as a scientist also showed Roberts that obtaining private funding required much work. Government funding might therefore fill in as a useful supplement to the private funding.

In the second phase, the WWII years—"the scientific emergency period" —saw the rise of the U.S. government as a dominant force in U.S. science. By necessity, the federal government entered into the creation and stimulation of scientific disciplines. Both space and atmospheric science benefited as disciplines from the need to generate a comprehensive understanding of the earth's atmosphere and near-earth space environment for national security purposes.[18] For meteorologists, both during the war and after, the role of government funding was not problematic. The federal sources had funded most meteorological enterprises for many years. Numerous younger theoretical and applied meteorologists came of age during WWII, and from early in their careers they understood the potentially large role the U.S. government could play in fostering their discipline.[19]

Roberts, still a young man during the war years, labored under the previous paradigm of private funding as best for science. He directly benefited from U.S. government funding as the military supported his work. Roberts came out of the war experience as a transition figure—still holding on to the old ideology of the importance of private financing, but also realizing the vital role of the U.S. government, including the U.S. military, in advancing questions relating to the sun–earth connection.

This amalgam of funding sources forms the basis of his approach to science funding in the third funding phase in his career—a time of "funding flux." In the decade between 1946 and October 1957, there were ebbs and flows in private and government support for Roberts's work. The ONR and Air Force provided much of his funding, but often his private funding from donations and foundations kept HAO going. Always sensitive to the vagaries of military funding and the problems of what he viewed as a troublesome bureaucracy, he never let go in this period of his desire to have a considerable amount of private funds to augment federal monies. Thus, he engaged in a nearly incessant drive for almost a decade to raise funds for his Boulder-based scientific efforts.

By the time of the IGY and *Sputnik*, Roberts was a mature man in his mid-forties. He made the full transition (seemingly effortlessly) to one who not only sought but welcomed full government funding of space and atmospheric science. Perhaps, as his colleague and friend Thomas Malone said, Roberts had shed some of his youthful ideology. Roberts realized that in funding science, "a better way" than private donations might exist.[20]

The "better way" became more tolerable to Roberts and others because of the increasing role beginning in the 1950s of the new non-military actors in organizing priorities and funding U.S. science—NSF, NASA, and NAS. The government owned and funded, but privately run (by UCAR), research facility NCAR served as a prime example.

This fourth and final phase lasted from 1957 until the mid-1960s—"the tsunami of science funding." Roberts happily rode this wave of funding in establishing NCAR. The funding created a science peak in Boulder, as Roberts simultaneously became a major voice in determining U.S. atmospheric science policy and funding priorities. Roberts had completed the transition to the "better way" of government funding for science.

Old habits die hard, and Roberts never totally relinquished his fundraising. Upon dedication of the multimillion-dollar, I. M. Pei–designed NCAR building in 1966, Roberts had another ceremony dedicating the "Ralph Damon Room" at NCAR. Roberts wanted to honor Damon, a former president of Trans World Airlines, and supporter of HAO, who died in 1956. Roberts, true to old form, solicited private funds to arrange for a room dedicated to his friend and HAO benefactor.[21]

All but forgotten now, even in Boulder and the scientific region that he helped to develop there, Roberts represents a key generation in U.S. science.[22] This generation grew up with the ethos of the "small science"—privately funded research given mostly to individual researchers. Many in that generation rapidly adjusted to these new realties and promises afforded by government-supported "big science." Even as the high tide of funding receded in the era of Vietnam and other tribulations of the late 1960s, the landscape of U.S. science had changed dramatically, both literally and figuratively, since the 1930s. Boulder was an integral part of that transformation.

As a transitional figure in this formative era of modern U.S. science, Roberts provided a major impetus in developing at least two major scientific disciplines—solar–terrestrial physics and atmospheric science. In doing this, he played an instrumental role in creating and sustaining a modern U.S. city of scientific knowledge production.

Roberts's career displays a strong continuity of interests, irrespective of sponsor or funding source. The core of these interests, although initially manifested in solar physics, centered on questions of the sun–earth connection. This continuity, spanning the four periods of science funding, shows how impervious his interests remained to the vagaries of his funding sources, whether government or private, military or non-military. The military, in particular, although a heavy funder of his work in Boulder until the late 1950s, never determined or manipulated his interests in any real sense. The Navy (ONR) and Air Force (via the Cambridge Research Center) essentially funded what he wanted to work on. Even during the national crisis of WWII, defense interests in radio propagation in effect provided the funds that allowed Roberts to complete his doctorate, initiated well before the war, while serving war needs. This seamless integration easily transitioned to the post-WWII period with the creation of HAO.

Boulder and Modern Regions of Scientific Knowledge

In some ways, Roberts's efforts in Boulder were not unlike oceanographer Roger Revelle's attempt to develop La Jolla, California, as a site for modern science. The Scripps Institute of Oceanography (SIO), established in 1903 as the Marine Biological Institute of San Diego, was even then a world leader in oceanographic science.[23] SIO's Revelle had envisioned the La Jolla area as a major center for oceanographic and associated scientific research in the 1950s. Among other things, he strongly advocated for the creation of a new University of California campus near SIO.[24]

Both Roberts and Revelle successfully exploited opportunities that the evolving U.S. science policy of the time offered them, but there were key differences between the two men's approaches to developing science in their respective locales. For one, Revelle saw the federal government, especially the military, as the guarantor of basic research in the postwar era and appealed directly to Cold War–era national security interests to advance his vision for scientific research in La Jolla.[25] He shared none of Roberts's desire to maintain some private funding in the years following WWII. Revelle's attitude on government funding was much closer to Menzel and Whipple's than Roberts's or Shapley's views.

Another major difference between Roberts and Revelle was that Roberts, as this research showed, fostered an environment (including local political support) for sun–earth research. Revelle appears to have had a much more institutional focus (on Scripps) for the nexus of development than a scientific quest analogous to Roberts's sun–earth interests.[26] This desire to make Boulder a center for sun–earth science transcended his specific work at the HAO, enabling all sorts of sun–earth connection and related research activities—labs, institutes, centers, and the like—to develop in Boulder in the 1950s and 1960s. Revelle's work, therefore, in advancing the development of science in La Jolla—mostly on creating UC San Diego as an extension of Scripps research— although similar in some respects to Roberts's in Boulder, represented in fundamental ways a different approach to the creation of a science region in this era.[27]

Boulder's story also differs from other ostensibly similar sites of American science such as Silicon Valley. Stanford University in the 1930s and 1940s, although not a powerhouse of physics or engineering on par with elite northeastern schools, had achieved notice. As early as the 1920s, the school's scientific research activities were on the way to becoming "moderately distinguished."[28] In the 1930s, William Hanson of the physics department invented the rhumbatron, an electronic device with a host of both research and commercial applications. This device attracted the attention of industries, including the Sperry Corporation and the Varian brothers, thereby helping to set the stage for the school's heavy interaction with industry and the government as years progressed.[29]

The rhumbatron also indicates another key difference between Boulder and Palo Alto. Much of Stanford's (and the area's) development centered on the production of useful technological devices with commercial and military application such as the rhumbatron. There was no analog in Palo Alto to the scientific quest that drove much of early Boulder science—the physics of the sun–earth connection. Much of Stanford's development centered on the engineering of things, and the physics research needed to support that work.[30] Stanford therefore had, eventually, a larger role in the military–industrial–academic complex of the 1950s than Boulder ever had (or wanted). Boulder rose as a center of scientific knowledge production; Palo Alto rose as a builder of things. It is not an accident then that the "Father of Silicon Valley," Stanford's Frederick Terman, was an electrical engineer. Roberts was a scientist. The differing initial paths Boulder and Palo Alto took represented a fundamentally different approach to knowledge generation and its uses.[31]

The Boulder story evokes Robert Merton's last work, *The Travels and Adventures of Serendipity*. The book is a history of the word "serendipity," including the use of the word in describing scientific findings. This word perhaps most accurately describes how Boulder developed, unexpectedly, into a world-class center of knowledge production. Serendipity as defined here is not just a random find—it results both from unexpected circumstances and those prepared to take advantage of those circumstances.[32] In broad outline, this is what occurred in Boulder. Because of this "serendipity factor," diverse attempts to create lists that outline criteria for the reproduction of science locales such as Boulder and Palo Alto are problematic, even chimerical at best, for they founder on the sharp rocks of the historical evidence. Each of the centers of knowledge production and components of the nation's military industrial complex had, as one study concludes, "unique" conditions.[33] As a result, these lists of criteria for growth may confuse symptom—a major university or heavy funding—with cause—a scientific pursuit, as in Boulder's case.

Boulder's development as a city of space and atmospheric science knowledge demonstrated the importance of many disparate groups of individuals in exploring and exploiting the diverse opportunities presented by the context of the early Cold War period. As was the situation with "Manchester science" in nineteenth century England, "Boulder science" of the twentieth century was a product of the manner in which local politics interacted with these emerging opportunities.[34] Thus, these opportunities for development—based on the sun–earth connection science—successfully created the conditions for a horizontally integrated community for scientific knowledge production to develop, expand, and then thrive.[35]

In 1623, Tommaso Campanella published *The City of the Sun*.[36] In this work he described a special place where learned individuals gathered together to imbibe in all human knowledge. The original focus of knowledge in Boulder was the sun and the sun–earth connection. In a very real sense,

Boulder became not simply a city of scientific knowledge production. The city became a modern, Cold War–era version of Campanella's vision for a scientific–academic utopia.

In the midst of the city on a hill, not unlike the Pei-designed NCAR building on the mesa, stood a "temple of astonishing design" dedicated to learning.[37] In the city, one learned more in a year than "ten or fifteen" anywhere else.[38] The complex and shifting context of twentieth-century American science combined with a compelling research idea and ambitious individuals to produce a city of scientific knowledge that in many ways evokes the spirit envisioned in the writings of Campanella and others long ago.

Walter Orr Roberts as a young man, no doubt doing something that was his passion. (UCAR Archives)

Climax Observatory in the 1960s. This is where Walter and Janet Roberts lived and worked from 1940 until 1946. The larger observatory building would not have been there at that time. (UCAR Archives)

Janet Roberts giving the Roberts's car a push up to Climax. (UCAR Archives)

It was not all work at Climax Observatory. Walter and Janet here are taking advantage of the plentiful snow at almost 7000 feet altitude. (UCAR Archives)

The Climax Observatory. (UCAR Archives)

Living in the Colorado mountains had its challenges; Janet Roberts doing laundry among the icicles. (UCAR Archives)

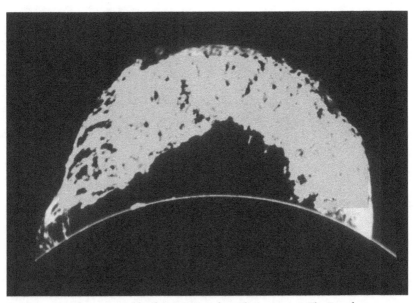

A solar prominence captured by the coronagraph at Climax in 1946. The French astronomer Lyot perfected the instrument that would allow for photographs such as this one. (UCAR Archives)

The first sign for the original Sacramento Peak Observatory in New Mexico. An Air Force funded project, its existence almost derailed the development of Boulder as a center for sun–earth connection research. Note the spelling of "Sacramento" on the sign. (UCAR Archives)

White Sands, the site of the US Army's V2 testing, as viewed from Sacramento Peak. The site was picked precisely due to the clear line of site seen here. (UCAR Archives)

The original coronagraph dome built at White Sands under the auspices of Harvard astronomers Donald Menzel and Roberts. (UCAR Archives)

Jack Evans, sitting on the truck. Along with Roberts, he argued that HAO not be closed so that "Sac Peak" could be better supported. Ironically, he later became the on-site director of the mountain top solar observatory for the Air Force. (UCAR Archives)

A meeting of the Harvard managers and the Air Force Cambridge Laboratory sponsors of the Sac Peak site. Donald Menzel is 5th from the left.

Alan Shapley of the Central Radio Propagation Lab. Working in close association with Roberts, he was essential in bringing more sun–earth science to Boulder, especially during the International Geophysical Year (1957–1958). (UCAR Archives)

Harvard's Donald Menzel, National Bureau of Standards (NBS) Director E. U. Condon, and Roberts. This photo was taken when the triumvirate worked together to bring the NBS's Central Radio Propagation Lab to Boulder. (*Boulder Daily Camera*)

THE PLAIN FACTS ABOUT

"PROSPERITY INSURANCE"

The Boulder Chamber of Commerce -- U. S. Bureau of Standards
Radio Laboratory Fund Campaign
STARTING APRIL 11th

WHAT IT IS...WHY...AND WHAT IT MEANS TO BOULDER

Pay Your Share of the Prosperity
Insurance Premium...TODAY!

Collect Your Share of the Dividends...TOMORROW!

The Boulder Chamber of Commerce made a direct appeal to businesses and citizens in flyers like this to buy "prosperity insurance." They would do this by contributing money that would go to purchase land to be donated to the U.S. Government for the relocation of the National Bureau of Standards Lab. Note the "answers" to the many questions asked. (Reprinted by permission of *Boulder Daily Camera*. Courtesy of Carnegie Library, Boulder.)

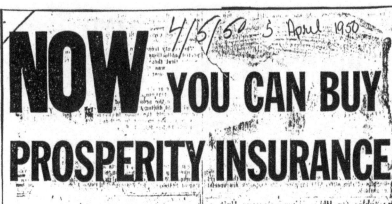

NOW YOU CAN BUY

PROSPERITY INSURANCE

You probably now have health, accident and life insurance. You also probably have car insurance, fire insurance, wind insurance — perhaps hail, theft, public liability and other forms of insurance protection.

BUT HOW ABOUT

YOUR BUSINESS ?

Will you have the customers and clients next year you have now? Will there be as much money circulating in the community next year? — and the next? — as there is today? Will your property be worth as much next year as it is today? Don't you wish you knew?
Of course no one can GUARANTEE what conditions will be like in the future but there is a way that you can HELP keep your business prosperous and that is

Help Bring the U.S. Bureau of Standards
Radio Laboratory to Boulder
This Multi-Million Dollar Project Means

BOULDER PROSPERITY
INSURANCE

With a Total Premium Cost of $70,000
It Pays An Annual Dividend of a $2,000,000 Payroll!
Can You Afford Not To Have This Protection?

Pay Your Share of the Premium — TODAY!
Collect Your Share of the Dividends — TOMORROW!

The Boulder Chamber of Commerce appealed directly to Boulder business owners to contribute money for the land that would be donated to the federal government. (Reprinted by permission of *Boulder Daily Camera*. Courtesy of Carnegie Library, Boulder.)

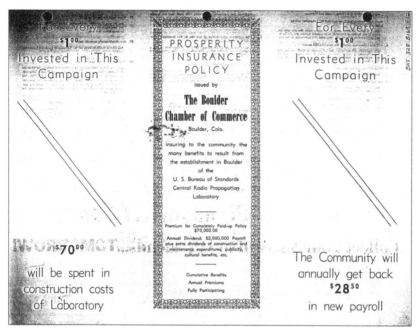

A faux "insurance policy" flyer distributed by the Chamber of Commerce in the fund drive campaign in 1950. (Carnegie Branch, Boulder Library)

President Eisenhower at the dedication of the NBS Boulder Labs in September, 1954. (NIST Archives)

A portrait of the large crowd that gathered for the dedication of the NBS Boulder Labs in September, 1954. Community involvement was essential in the move to bring the labs to Boulder. (NIST Archives)

Sydney Chapman, known to many for his bon homie. Besides being a major figure in sun–earth connection studies, he was also a close associate of Walter Roberts in advancing sun–earth connection science. (UCAR Archives)

ELEVEN REASONS WHY
YOU SHOULD VOTE FOR THE NCAR
(NATIONAL CENTER FOR ATMOSPHERIC RESEARCH)
BLUE LINE AMENDMENT TOMORROW!

1. Only by supplying water to the NCAR site can the City of Boulder annex this area and thus gain some control over its development.

2. If the City does not supply water for NCAR, development of another water source may open the area south of the site to private development, over which the City would have no control what so ever.

3. NCAR has given complete assurances that the natural beauty of the area will be preserved and that the Mesa Trail will be protected and kept open.

4. The National Center for Atmospheric Research will add greatly to Boulder's importance as a research center and will result in favorable world wide prestige.

5. NCAR will benefit Boulder's economy by providing depression proof employment.

6. NCAR will pay the City for services provided.

7. NCAR's water requirements are modest, approximately that used by 25 homes.

8. NCAR will be of great value to the future of the University of Colorado.

9. NCAR will protect and preserve the spirit and objective of the Blue Line.

10. An overwhelming vote in favor of the amendment will assure the National Science Foundation that their judgment was correct in selecting Boulder over numerous other locations bidding for NCAR.

11. And an overwhelming vote in favor of the amendment is of the utmost importance to encourage the Colorado State Legislature to appropriate the necessary funds for the purchase of the site.

For these reasons the following Boulder Citizens have endorsed the passage of the Blue Line Amendment and have volunteered to the Citizens Committee the use of their names in support of a favorable vote.

GROUPS, CLUBS & ORGANIZATIONS

have by formal action gone on record in support of the passage of the Blue Line NCAR Amendment.

BOULDER BOARD OF REALTORS
BOULDER CHAMBER OF COMMERCE
BOULDER CITY COUNCIL
BOULDER COUNTY COMMISSIONERS
BOULDER INSURORS ASSOCIATION
BOULDER JOURNEYMEN BARBER'S UNION
BOULDER ROTARY CLUB Board of Directors

BOULDER LIONS CLUB
BUSINESS AND PROFESSIONAL WOMEN'S CLUB
COSMOPOLITAN INTERNATIONAL OF BOULDER
JAYCEE JEMS
JUNIOR CHAMBER OF COMMERCE
KIWANIS CLUB OF BOULDER
PLAN BOULDER
SOROPTIMIST CLUB
BOULDER OPTIMIST CLUB

THE ABOVE ARE ONLY A FEW OF THE BOULDER CITIZENS WHO URGE YOU TO

Vote FOR Amendment No. 1 - Tomorrow

Dr. Walter Orr Roberts and City Manager E. Robert Turner will answer questions called into them concerning Amendment No. 1 this evening on KBOL. All Boulder Residents are urged to listen to this program.

CALL HI 2-2242   **KBOL**   CALL HI 2-2242
7:00 to 8:00 p.m. TONIGHT

CITIZENS COMMITTEE FOR THE NCAR BLUE LINE AMENDMENT
Ben Beeson, Chairman

A flyer appealing to Boulder voters to amend the "Blue Line" for NCAR. This is reminiscent of the flyers created in 1950 that asked Boulder citizens to donate money for the purchase of land for the NBS. (Carnegie Branch, Boulder Library)

Constructing the iconic, I. M. Pei designed, National Center of Atmospheric Research building. (UCAR Archives)

What the NCAR building might have looked like. This is a model of an alternate and somewhat futuristic proposal of the NCAR campus that was not selected. (UCAR Archives)

NCAR today, with the well-known "Flatirons" as backdrop. The almost monastic-like architecture of the building reflects Walt Robert's vision of the ideal space for scientists to interact. (UCAR Archives)

Walter Roberts and I. M. Pei in front of their creation. (UCAR Archives)

The Walt Roberts "Wanted Poster." Although done in a spirit of fun by NCAR associates, this poster demonstrates the skepticism many atmospheric scientists had about Roberts's desire to connect the sun and earth via weather prediction. (UCAR Archives)

Notes

Notes to Chapter 1

1. Job 38:33. This is the inscription on the U.S. National Academy of Sciences' International Geophysical Year poster on "The Sun & Earth." National Academy of Sciences, "The Sun & Earth," accessed June 8, 2015, http://www.nasonline.org/about-nas/history/archives/milestones-in-NAS-history/international-geophysical-year-files/igy-picture-galleries/picture-gallery-images/sunandearth_large.jpg.

2. Christopher R. Henke and Thomas F. Gieryn, "Sites of Scientific Practice: The Enduring Importance of Place," in *The Handbook of Science and Technology Studies*, ed. Edward J. Hackett et al. (Cambridge, MA: MIT Press, 2008), 355.

3. For a short discussion of Tommaso's idea, see Robert Kargon, Stuart W. Leslie, and Erica Schoenberger, "Far beyond Big Science: Science Regions and the Organization of Research and Development," in *Big Science: The Growth of Large-Scale Research*, eds. Peter Galison and Bruce Hevly (Stanford: Stanford University Press, 1992), 336.

4. Jean Claude Pecker, interview by Spenser Weart, September 11, 1976, OH 366, transcript, Niels Bohr Library & Archives, AIP, College Park, MD, 38. The original quote is in French: "Boulder, c'est une grande universite, maintenant, ce que j'appelle, moi, 'Astroboulder.' C'est beaucoup de choses."

5. Carol Knight, "Ike Liked the Labs: How a 'Scientific Siberia' Became the Hub of U.S. Atmospheric Research," NOAA Research, accessed September 15, 2005, http://www.oar.noaa.gov/spotlight/archive/spot_ike.html (site discontinued).

6. This number does not include over 35,000 students at the University of Colorado Boulder.

7. Paul Danish, "The People, Yes and No," *Coloradan*, March 2008, 12, accessed April 30, 2015, https://alumni.colorado.edu/wp-content/uploads/coloradan/2008_03/10_news.pdf.

8. "The 10 Smartest Cities in America," *MarketWatch*, January 31, 2015, accessed June 8, 2015, http://www.marketwatch.com/story/the-10-smartest-cities-in-america-2015-01-02.

9. ESSA became NOAA in 1970.

10. Barbara B. Poppe and Kristen P. Jordan, *Sentinels of the Sun: Forecasting Space Weather* (Boulder, CO: Johnson Books, 2006). Read chapters 5 and 6 in particular for a history of space weather services.

11. "General Information," National Institute of Standards and Technology, Boulder Laboratories, accessed October 15, 2008, http://www.boulder.nist.gov/geninfo.htm (site discontinued).

12. For information on JILA, see https://jila.colorado.edu/about/about-jila, accessed June 8, 2015.

13. "Satellite spies on solar flare cycles," *Coloradan*, March 2008, 20, accessed April 30, 2015, https://alumni.colorado.edu/wp-content/uploads/coloradan/2008_03/10_news.pdf.

14. At the time of the author's graduate student days in this program during the early 1980s, the program had the unwieldy title of "Astrophysical, Planetary, and Atmospheric Sciences."

15. Karl Hufbauer, *Exploring the Sun: Solar Science since Galileo* (Baltimore: Johns Hopkins University Press, 1991), 169–172.

16. Margaret Pugh O'Mara, *Cities of Knowledge: Cold War Science and the Search for the Next Silicon Valley* (Princeton: Princeton University Press, 2005), 1.

17. Ann Markusen, Scott Campbell, Peter Hall, and Sabina Detirick, eds., *The Rise of the Gunbelt: The Military Remapping of Industrial America* (New York: Oxford University Press, 1991), 175–176.

18. O'Mara, *Cities of Knowledge*, 227; Rebecca S. Lowen, *Creating the Cold War University* (Berkeley: University of California Press, 1997); Stuart W. Leslie, *The Cold War and American Science: The Military-Industrial-Academic Complex at MIT and Stanford* (New York: Columbia University Press, 1993); Paul E. Ceruzzi, *Internet Alley: High Technology in Tysons Corners, 1945–2005* (Cambridge, MA: MIT Press, 2008), 14.

19. Daniel J. Kevles, *The Physicists: The History of a Scientific Community in Modern America* (New York: Vintage Books, 1979), 192.

20. Leslie, *The Cold War*, 2. For an example of a study of another region, see Ronald Rainger, "Constructing a Landscape for Postwar Science: Roger Revelle, The Scripps Institution and The University of California, San Diego," *Minerva* 39, no. 3 (2001): 327–352, doi:10.1023/A:1017952811593.

21. David H. DeVorkin, *Science with a Vengeance: How the Military Created the US Space Sciences after World War II* (New York: Springer-Verlag, 1992); Ronald E. Doel, "Constituting the Postwar Earth Sciences: The Military's Influence on the Environmental Sciences in the USA after 1945," *Social Studies of Science* 33, no. 5 (2003): 635–666, doi:10.1177/0306312703335002.

22. Vannevar Bush, interview by Eric Hodgin, Reel 4-B, 259, 1964, MC-143, Institute Archives and Special Collections, MIT, at AIP/Center for the History of Physics, College Park, MD.

23. Michael A. Dennis, "Our First Line of Defense: Two University Laboratories in the Postwar American State," *Isis* 85, no. 3 (1994): 453.

24. The only exceptions that may contradict this claim are the establishment of the Atomic Energy Commission (AEC) Cryogenics Laboratory in Boulder in 1954 (in conjunction with the NBS move there) and the creation of the AEC Rocky Flats nuclear weapons plant. However, neither facility played a major role in the story of science in Boulder per se, or in the city's ascending trajectory as a place for sun–earth science in the 1950s. As an indication of the tangential role that the plant played in the city's history, Boulder does not even rate a mention in the definitive account of the plant's early history. See Len Ackland, *Making a Real Killing: Rocky Flats and the Nuclear West* (Albuquerque: University of New Mexico Press, 1999). As indicated by this work's title, the Rocky Flats narrative fits more into the nuclearization of the Rocky Mountain West and the rise of the "Gunbelt" in the post-WWII United States. In some ways the weapons plant actually emphasizes the uniqueness of what occurred in Boulder in the 1950s, for what occurred in Boulder is not analogous to the developments relating to the "nuclear west."

25. See Robert K. Merton, "Science and Democratic Social Structure," in *Social Theory and Social Structure*, ed. Robert K. Merton (New York: Free Press, 1968).

26. "Dual use" is the term often applied to knowledge, techniques, and devices useful to both the military and non-military components of society.

27. For a discussion of the centralized nature of the planning for Los Alamos, see Robert S. Norris, *Racing for the Bomb: General Leslie R. Groves, The Manhattan Project's Indispensable Man* (South Royalton: Steerforth Press, 2002), 231–253. Also see Carl Abbott, "Building the Atomic Cities: Richland, Los Alamos and the American Planning Language," in *The Atomic West*, eds. Bruce Hevly and John M. Findlay (Seattle: University of Washington Press, 1998), 98–115. For a brief discussion of the centrally planned Soviet science cities, see Zhores A. Medvedev, *Soviet Science* (New York: W.W. Norton, 1978), 74–75.

28. Markusen, "Space Mountain: Generals and Boosters Build Colorado Springs," *The Rise of the Gunbelt*, 174–210.

29. Robert Rosenberg, *Boulder, Colorado: Development of a Local Scientific Community: 1939–1960* (unpublished manuscript, received at NCAR Archives October 1982), 77. Rosenberg's observation was an important beginning in understanding what transpired in Boulder.

30. AnnaLee Saxenian, *Regional Advantage: Culture and Completion in Silicon Valley and Route 128* (Cambridge, MA: Harvard University Press, 1994). Ceruzzi in Internet Alley also considers social dynamics in the rise of technological regions.

31. Saxenian, *Regional Advantage*, 2.

32. Christophe Lécuyer, *Making Silicon Valley: Innovation and the Growth of Silicon Valley* (Cambridge, MA: MIT Press, 2007), 4–5. Lécuyer refers to the work of nineteenth-century British economist Alfred Marshall. For Marshall's original discussion on the concentration of specialized industries in a particular location, see his *Principles of Economics* (London: MacMillan and Company, 1891), 326–336.

33. Horizontal integration invokes philosopher of science Michael Polanyi's idea of a "republic of science," where independent scientific activities in effect self-coordinate via an "invisible hand" by simply being "in sight" of each other. Michael Polanyi, "The Republic of Science: Its Political and Economic Theory," *Minerva* 38, no. 1 (2000): 1–3, doi:10.1023/A:1026591624255. Although Polanyi wrote metaphorically in describing the mechanisms of scientific communication, in Boulder's case the condition of research sites being in sight of one another almost literally applies.

34. Kargon et al., "Far beyond Big Science," 335–336. Horizontal integration may also be the basis of large-scale "big science" activities as well.

35. W. Patrick McCray, *Giant Telescopes: Astronomical Ambition and the Promise of Technology* (Cambridge, MA: Harvard University Press, 2004), 34–42.

36. Roger L. Geiger, "Science, Universities, and National Defense, 1945–1970," *Osiris, Second Series* 7 (1992), 26–48.

37. William Shakespeare, *Julius Caesar*, act 4, scene 3, lines 218–219. Brutus speaks.

38. Markusen et al., *The Rise of the Gunbelt*, 5–6, and throughout.

39. Ronald E. Doel, *Solar System Astronomy in America: Communities, Patronage, and Interdisciplinary Research, 1920–1960* (New York: Cambridge University Press, 1996), 198.

40. For example, as one of the four scientific themes in its current Office of Space Science, NASA now has a "Sun–Earth Connection" program that encompasses some of these studies (see http://sec.gsfc.nasa.gov/). NASA also has a "Sun–Earth Connection Education Forum" (http://sunearth.gsfc.nasa.gov/; site discontinued).

41. Many use 1957 and the launch of *Sputnik* as the advent of the Space Age, but the term is sometimes used more broadly and therefore covers a broader time period.

42. "Earth Sciences in the Cold War." Special issue, *Social Studies of Science* 33, no. 5 (2003).

43. Doel, "Constituting the Postwar Earth Sciences," 635–666.

44. Ronald E. Doel, "Earth Sciences and Geophysics" in *Companion to Science in the Twentieth Century*, eds. John Krige and Dominique Pestre (London: Routledge, 2003), 391–416.

45. Michael A. Dennis, "Earthly Matters: On the Cold War and the Earth Sciences," *Social Studies of Science* 33, no. 5 (2003): 809–819.

46. Rosenberg, *Boulder, Colorado*, 42.

47. David N. Spires, "Walter Roberts and the Development of Boulder's Areospace Community," *Quest: The History of Spaceflight Quarterly* 6, no. 4 (1998): 5–14. This article served as one of the principle motivations and foundations for this research.

48. Elizabeth Lynn Hallgren, *The University Corporation of Atmospheric Research and The National Center for Atmospheric Research, 1960–1970: An Institutional History* (Boulder, CO: UCAR, 1974).

49. George T. Mazuzan, "Up, Up, and Away: The Reinvigoration of Meteorology in the United States: 1958–1962," *Bulletin of the American Meteorological Society* 69, no. 10 (1988): 1152–1163.

50. Thomas J. Bogdan, "Donald Menzel and the Beginnings of the High Altitude Observatory," *Journal of the History of Astronomy* 33 (2002): 159. In addition to HAO, his comment refers to the Sacramento Peak Observatory in New Mexico and the Harvard Radio Astronomy Observatory at Fort Davis, Texas.

51. Walter Sullivan, *Assault on the Unknown: The International Geophysical Year* (New York: McGraw Hill, 1961).

52. Ceruzzi, *Internet Alley*, 165.

Notes to Chapter 2

1. Stuart Clark, *The Sun Kings: The Unexpected Tragedy of Richard Carrington and the Tale of How Modern Astronomy Began* (Princeton: Princeton University Press, 2007), 37.

2. Norman Lockyer, "Progress in Astronomy during the Nineteenth Century," *Smithsonian Report for 1900* (Washington, D.C.: Government Printing Office, 1901): 145.

3. David DeVorkin, "Defending a Dream: Charles Greeley Abbot's Years at the Smithsonian," *Journal of the History of Astronomy* 21 (1990): 121–135.

4. Hufbauer, *Exploring the Sun*, 115.

5. DeVorkin, *Science with a Vengeance*, 231. This is not to say serious solar research did not exist at the time. The point is that solar research still stood apart from the primary interests of the astronomers such as galaxies, stellar phenomena, and similar studies.

6. Walter Roberts, interview by David DeVorkin, July 26 and 28, 1983, transcript, Niels Bohr Library & Archives, AIP, College Park, MD, 52.

7. O'Mara, *Cities of Knowledge*, 107.

8. In the case of Terman and Draper, see Leslie, *The Cold War*, for extended discussions of their importance in creating major sites of technological production. See O'Mara in *Cities of Knowledge*, chapter 3, for a similar discussion of Terman's importance to the region.

9. Markusen et al., *The Rise of the Gunbelt*, 178.

10. Kargon et al., "Far beyond Big Science," 353.

11. The summer of 1936, for example, holds the twentieth-century record for the highest average summer temperatures, with 1934 running a close second. National Environmental Satellite, Data and Information Service, "2002 Summer Hot, Dry across Much of United States; Upper Mid-West Experiences Wetter-than-Average Summer," accessed June 8, 2015, http://www.publicaffairs.noaa.gov/releases2002/sep02/noaao2119.html.

12. Roberts, interview by DeVorkin. In claiming this, Roberts reveals himself as having in effect a Baconian view that the true importance of science knowledge rests in its utility. For a discussion of how this idea of "operationalism" might have shaped modern science, see Peter Dear, *Revolutionizing the Sciences: European Knowledge and Its Ambitions, 1500–1700* (Princeton, NJ: Princeton University Press, 2001).

13. Roberts, interview by DeVorkin, 1–3.

14. Roberts, interview by DeVorkin, 12.

15. Roberts, interview by DeVorkin, 16–17.

16. Roberts, interview by DeVorkin, 17.

17. Donald H. Menzel, audiotape of remarks at High Altitude Observatory Thirtieth Anniversary Celebration, October 2, 1970, Tape 1, OHP 90-91, UCAR Archives, Boulder, CO.

18. Roberts, interview by DeVorkin, 17.

19. Dorrit Hoffleit, "Shapley, Harlow," in *History of Astronomy: An Encyclopedia*, ed. John Lankford (New York: Garland Publishing, Inc., 1997), 461.

20. David DeVorkin, "Donald Menzel." *New Dictionary of Scientific Biography* (Detroit: Charles Scribner's Sons, 2008), 5, 110–111.

21. DeVorkin, "Donald Menzel," 111.

22. Doel, *Solar System Astronomy*, 23.

23. McCray, *Giant Telescopes*, 14.

24. Hoffleit, "Shapley, Harlow," 461. Ant behavior, in particular, fascinated the astronomer. He concluded, for example, that ants move faster as air temperature increases. For an extended discussion of Shapley's interest and observation of ants, see Harlow Shapley, *Through Rugged Ways to the Stars: The Reminiscences of an Astronomer* (New York: Charles Scribner's Sons, 1969), 65–72. When maintenance personnel at Mt. Wilson observatory used blow torches to rid the grounds of ants, Shapley thought them guilty of "genocide" (66).

25. John Lankford, *American Astronomy: Community, Careers, and Power, 1859–1940* (Chicago: University of Chicago Press, 1997), 370.

26. Doel, *Solar System Astronomy*, 27.

27. McCray, *Giant Telescopes*, 35.

28. A polarimeter works on the same principle that polarized sunglasses do.

29. Hufbauer, *Exploring the Sun*, 92.

30. Hufbauer, *Exploring the Sun*, 93–94.

31. Walter Roberts, "Research in the Rockies," *Appalachia*, June (1946): 33.

32. Bogdan, "Donald Menzel," 159.

33. Bogdan, "Donald Menzel," 161. Shapley was a complex personality—he could be both petty in one instance and magnanimous in another. His approach therefore engendered contradictory feelings by his subordinates. Roberts indicated these ambivalent feelings when saying that although Shapley was often guilty of "dirty tricks" in running HAO, Roberts still had great personal affection for him. Roberts, interview by DeVorkin, 85.

34. Lankford, *American Astronomy*, 209; David DeVorkin, "Who Speaks for Astronomy? How Astronomers Responded to Government Funding after World War II," *Historical Studies in the Physical Sciences*, 31, no. 1 (2000): 59.

35. Lankford, *American Astronomy*, 206. The Rockefeller Foundation, for example, funded George E. Hale's Palomar Observatory and businessman John D. Hooker provided money for his Mt. Wilson 100 inch telescope. For a discussion of the challenges of fundraising in the pre-WWII era, see Helen Wright, *Explorer of the Universe: A Biography of George Ellery Hale* (New York: AIP Press, 1994). Private donors funded all of Hale's astronomical research.

36. Lankford, *American Astronomy*, 189, 206.

37. DeVorkin, "Who Speaks for Astronomy?" 59.

38. As the years passed, the observatory would have an amazingly eclectic array of donors. Chapter 4 presents in detail Roberts's efforts in soliciting from and working with these donors.

39. Leo Goldberg, interview by Spencer Weart, May 16, 1978, transcript, Niels Bohr Library & Archives, AIP, College Park, MD, 33.

40. Lankford, *American Astronomy*, 188–189.

41. Lankford, *American Astronomy*, 189–190.

42. Hufbauer, *Exploring the Sun*, 89.

43. Clark, *The Sun Kings*, 59.

44. Clark, *The Sun Kings*, 56.

45. DeVorkin, "Defending a Dream," 121–135. This article presents in detail Abbot's ideas and the controversy surrounding them.

46. DeVorkin, "Defending a Dream," 129–130.

47. Walter Roberts, interview by DeVorkin, 25; Bogdan, "Donald Menzel," 161.

48. Bogdan, "Donald Menzel," 161.

49. DeVorkin, "Donald Menzel," 113.

50. DeVorkin, "Donald Menzel," 113.

51. Bogdan, "Donald Menzel," 161.

52. David Roberts, *On the Ridge between Life and Death* (New York: Simon and Schuster, 2005), 239. Walter Roberts's Harvard journal is in the possession of his family and not publicly available.

53. Roberts, interview by DeVorkin, 25.

54. DeVorkin, "Donald Menzel," 113.

55. Roberts, interview by DeVorkin, 28.

56. Bogdan, "Donald Menzel," 169.

57. Walter Roberts, interview by Stephen Gassaway, May 13, 1987, OH 0519, Carnegie Library, Boulder, CO.

58. Walter Roberts, interview by William H. Jackson, High Altitude Observatory Thirtieth Anniversary Celebration, October 2, 1970, OHP 90-91, UCAR Archives, Boulder, CO. This is a less formal discussion between Roberts and Jackson.

59. Bogdan, "Donald Menzel," 169.

60. Roberts, interview by Jackson.

61. Bogdan, "Donald Menzel," 170.

62. Bogdan, "Donald Menzel," 170–171.

63. Roberts, "Research in the Rockies," 34; Bogdan, "Donald Menzel," 178.

64. Harriet Barker Crowe, John Firor, Betty O'Lear, Ed Wolf, and Mary Wolf, *Remembering Walter Roberts* (Boulder, CO: UCAR, 1991), 13.

65. D. Roberts, *On the Ridge*, 18.

66. Gino Segre, *Faust in Copenhagen: A Struggle for the Soul of Physics* (New York: Viking Press, 2007), 95. As of the 1970s, based on this author's early experience in physics, this notion still existed in the physics community.

67. Bogdan, "Donald Menzel," 181.

68. Bogdan, "Donald Menzel," 182.

69. The east limb is the left side of the solar disk as viewed from earth.

70. Hufbauer, *Exploring the Sun*, 90.

71. Walter Roberts, "Questionnaire for History of Modern Astronomy," (AIP/CHP, n.d.): 1.

72. Hufbauer, *Exploring the Sun*, 119.

73. Hufbauer, *Exploring the Sun*, 89.

74. Roberts, interview by DeVorkin, 42.

75. Hufbauer, *Exploring the Sun*, 121.

76. Bogdan, "Donald Menzel," 182.

77. Roberts, "Research in the Rockies," 36.

78. Bogdan, "Donald Menzel," 182.

79. Roberts, "Research in the Rockies," 36–37.

80. In a dramatic counterpoint to Roberts's sheltered, scientific life in the Rockies, his younger brother, Stuart, died flying a navy bomber during the battle for Saipan in the summer of 1944.

81. For a description of how the Manhattan Project could get any support it wanted during the war, see Norris, *Racing for the Bomb*, 317.

82. Roberts, interview by DeVorkin, 42.

83. Roberts, interview by DeVorkin, 42; Roberts, AIP Questionnaire, 1.

84. Janet Roberts, interview by Shirley Steele, May 4, 2004, OH 1202V, transcript, Oral History Program, Carnegie Library, Boulder, CO, 2.

85. Roberts, "Research in the Rockies," 31.

86. Roberts, AIP Questionnaire, 1; Hufbauer, *Exploring the Sun*, 120.

87. Roberts, interview by DeVorkin, 42. Roberts errs somewhat in his recollection in this interview. He sent the discovery paper for "spicules" to the *Astrophysical Journal* in November 1944. Military authorities must have given him publishing approval, as he discussed the Climax operation in this article. See note 146. This also indicates the military did not view sun–earth connection science as a subject they wished to classify indefinitely.

88. Hufbauer, *Exploring the Sun*, 120.

89. Hufbauer, *Exploring the Sun*, 125.

90. Roberts, interview by DeVorkin, 42.

91. DeVorkin, "Donald Menzel," 114.

92. Walter Roberts, "A Preliminary Report on Chromospheric Spicules of Extremely Short Lifetime," *Astrophysical Journal* 101 (1945): 136.

93. Roberts, interview by DeVorkin, 47.

94. Roberts, "A Preliminary Report," 136.

Notes to Chapter 3

1. Hufbauer, *Exploring the Sun*, 129.

2. DeVorkin, "Donald Menzel," 114–115.

3. Doel, *Solar System Astronomy*, 68–77.

4. Geiger, "Science, Universities, and National Defense," 41.

5. Patrick McCray, "The Contentious Role of a National Observatory," *Physics Today*, 56, no. 10 (2003): 55–61.

6. Doel, *Solar System Astronomy*, 73.

7. Hufbauer, *Exploring the Sun*, 119.

8. DeVorkin, "Who Speaks for Astronomy?" 72.

9. Marcia Bartusiak, *The Day We Found the Universe* (New York: Pantheon, 2009), 166; Roberts, interview by DeVorkin, 82. Bartusiak's work discusses Shapley's complex personality, in addition to his management style while directing HCO. He somehow produced simultaneously great affection and disdain among his subordinates, as Roberts's words indicate. Roberts also told DeVorkin that everything he learned about "running a happy ship" he learned from Shapley.

10. Roberts, interview by DeVorkin, 87.

11. Harlow Shapley to Donald Menzel, 23 October 1944, fol. correspondence, 1942 P-Z, 1943, 1944, HUG 4567.5 DHM, Harvard University Archives (hereafter cited as HUA), Cambridge, MA. I am much indebted to Thomas Bogdan for generously sharing his files on the early history of HAO with me.

12. Donald Menzel to Harlow Shapley, 27 October 1944, HUA. Penrose was the same Harvard alum and Colorado Springs booster that Menzel approached in 1939, and who had rebuffed him at that time.

13. Harlow Shapley to James B. Conant, 26 March 1945, fol. Harlow Shapley Director 1940–1950, Collection UAV 630.37 HCO, HUA.

14. The Research Corporation, founded in 1912 as philanthropic enterprise to sponsor innovation, was a foundation and not a bona fide legal corporation, despite its name.

15. Robert P. Crease, *Making Physics: A Biography of Brookhaven National Laboratory, 1946–1972* (Chicago: University of Chicago Press, 1999), 28. For a discussion on Brookhaven's origins, see also Allan A. Needell, "Nuclear Reactors and the Founding of Brookhaven National Laboratory," *Historical Studies in the Physical Sciences* 14, no. 1 (1984), 93–122.

16. Regents Minutes, 21 September 1945, University of Colorado Archives (hereafter cited as CUA), Boulder, CO.

17. Regents Minutes, 26 October 1945, CUA.

18. Albert A. Bartlett and Jack J. Kraushaar, *A History of the Physics Department of the University of Colorado at Boulder, 1876–2001* (Boulder: Department of Physics, University of Colorado, 2002), 11–13.

19. Roberts, interview by DeVorkin, 79.

20. CU News Release, 14 April 1946, fol. 2, NCAR Collection, Carnegie Library, Boulder, CO.

21. Roberts, interview by DeVorkin, 46. As charming as this story is, however, the record shows that Shapley had used the term to describe the new Colorado operation a year or so before, so Roberts's recollection is flawed in this regard. See, for example, Harlow Shapley to James Conant, 17 May 1945, Correspondence Papers, 1939–1945, UAV, 630.22.5 HCO, HUA.

22. Hallgren, *The University Corporation*, 57.

23. Roberts, interview by Jackson.

24. Hallgren, *The University Corporation*, 58.

25. DeVorkin "Donald Menzel," 114–115; McCray, *Giant Telescopes*, 34.

26. DeVorkin, "Donald Menzel," 114.

27. D. Roberts, *On the Ridge*, 18.

28. Roberts, interview by DeVorkin, 91.

29. Roberts, interview by Gassaway. Harvard began a similar effort to discuss the interrelationships between science and society about this time. For some of Roberts's views, see his series of lectures on science and society given at Claremont College published as Walter Orr Roberts, *A View of Century 21* (Claremont, CA: The Claremont Colleges, 1969).

30. Bartlett and Kraushaar, *A History*, 13–15.

31. Roberts, interview by Gassaway.

32. Rosenberg, *Boulder, Colorado*, 23.

33. William E. Davis, *Glory Colorado: A History of the University of Colorado, 1858–1963* (Boulder, CO: Pruett Press, 1965), 6.

34. Davis, *Glory Colorado*, 8.

35. Rosenberg, *Boulder, Colorado*, 33.

36. Rosenberg, *Boulder, Colorado*, 34. This was the son of the famed nineteenth-

century landscape designer. Among other places, Olmstead designed New York's Central Park.

37. The population increased from 9,500 to about 12,000 citizens.

38. Rosenberg, *Boulder, Colorado*, 37; U.S. Census and Public Facilities Plan and Capital Improvements Program, 1963–1985, City of Boulder (1963).

39. Davis, *Glory Colorado*, 589. Stearns, a Republican, resigned in 1952 because of controversies with the school's regents in defending faculty over academic freedom issues and from charges of communist affiliations, 577–588.

40. A letter from General Leslie R. Groves, head of the Manhattan Project, in October 1945 to Stearns attests to the contacts he made. Groves thanks him for his "important contributions and splendid cooperation" on the "atom bomb project" while serving in the Pacific as an operations analyst. L. R. Groves to Robert L. Stearns, 5 October 1945, CA/Pres Off I-100-5, Presidents' Office Files, CUA. What role Stearns played in the Manhattan Project remains unknown.

41. Davis, *Glory Colorado*, 589.

42. Hufbauer, *Exploring the Sun*, 132.

43. Ruth P. Liebowitz, "Donald Menzel and the Creation of the Sacramento Peak Observatory," *Journal for the History of Astronomy* 33 (2002): 195.

44. See DeVorkin, *Science with a Vengeance*, for a complete discussion of the U.S. military-sponsored V-2 research program.

45. Liebowitz, "Donald Menzel," 196.

46. Doel, *Solar System Astronomy*, 71–72; DeVorkin, *Science with a Vengeance*, 278. Whipple also considered Climax as a site for his meteor research, but ruled it out because of the number of cloudy days. Although most noted for atomic work, there exists to this day a large amount of both civilian and military scientific activity in the state. How these activities developed so far from the eastern scientific establishment in the decades following WWII is a question worthy of future study.

47. Roberts, interview by DeVorkin, 59.

48. Liebowitz, "Donald Menzel," 198.

49. Roberts, interview by DeVorkin, 59.

50. Walter Roberts to Harlow Shapley, 17 July 1947, fol. Harlow Shapley thru June 1948, box 1940–1957 s-u, Walter Orr Roberts Collection (hereafter WOR)/CUA, 1.

51. Roberts to Shapley, 17 July 1947, 1.

52. Harlow Shapley to Walter Roberts, 5 November 1947, fol. Harlow Shapley thru June 1948, box 1940–1957 s-u, WOR/CUA.

53. Roberts to Shapley, 17 July, 1947, 1 (emphasis in original).

54. Roberts to Shapley, July 17, 1947, 2.

55. Paul Forman, "Behind Quantum Electronics: National Security as a Basis for Physical Research in the United States," *Historical Studies in the Physical and Biological Sciences* 18, no. 1 (1987): 216.

56. Harlow Shapley to Walter Roberts, 25 July 1947, fol. Harlow Shapley thru June 1948, box 1940–57, s-u, WOR/CUA, enclosure (hereafter Shapley fol.).

57. DeVorkin, "Who Speaks for Astronomy?" 66, 70.

58. Shapley to Roberts, 25 July 1947, enclosure.

59. Forman, "Behind Quantum Electronics," 206.

60. James Hershberg, *James B. Conant: Harvard to Hiroshima and the Making of the Nuclear Age* (New York: Knopf, 1993), 446–448.

61. The open nature of scientific exchange has fostered this scientific ethos, sometimes also referred to as "cosmopolitanism." Shapley himself once stated a "scientist's blood" is by nature "cosmopolitan." Steven Shapin, *The Scientific Life: A Moral History of a Late Modern Vocation* (Chicago: University of Chicago Press, 2008), 67–68.

62. Jessica Wang, *American Science in an Age of Anxiety: Scientists, Anticommunism, and the Cold War* (Chapel Hill: University of North Carolina Press, 1999), 122–123.

63. Wang, *American Science*, 125.

64. Roberts, interview by DeVorkin, 86.

65. Hershberg, *James B. Conant*, 617.

66. DeVorkin, "Donald Menzel," 116.

67. Shapley to Roberts, 25 July 1947, enclosure.

68. Shapley to Roberts, 25 July 1947, enclosure.

69. Walter Roberts to Harlow Shapley, 13 August 1947, Shapley fol., 1.

70. Alamogordo, New Mexico, was the nearest city to the general area in the Sacramento Mountains where HCO expected to put the new facility.

71. Harlow Shapley to Walter Roberts, 21 August 1947, Shapley fol., 1.

72. Shapley to Roberts, 21 August 1947, 2.

73. Harlow Shapley to Walter Roberts, 19 May 1946, fol. Harlow Shapley, 1946, box 13, WOR/CUA, 2. Alamogordo is the nearest city of note to the Sacramento Peak site, and documents of the time often use the names interchangeably.

74. Shapley to Roberts, 21 August 1947, 2.

75. Walter Roberts to Harlow Shapley, 26 August 1947, fol. Harlow Shapley thru June 1948, box 1940–1957 s-u, WOR/CUA. 1.

76. As it turned out, the Air Force–funded facility on Sacramento Peak quickly became self-justifying, so Roberts's apparent concern, real enough at the time, did not develop as he thought it would.

77. Roberts to Shapley, 26 August 1947, 2.

78. Roberts to Shapley, August 26, 1947, 3.

79. Roberts to Shapley, August 26, 1947, 2.

80. Harlow Shapley, "Saturday Night Afterthoughts Relative to Alamogordo," 13 December 1947, memo, fol. Harlow Shapley thru June 1948, box 1940–1957 s-u, WOR/CUA, 1.

81. David Bushnell, *The Sacramento Peak Observatory* (Washington, D.C.: Office of Aerospace Research, 1962), 5.

82. Liebowitz, "Donald Menzel," 200.

83. Despite Roberts's claims to Shapley that living at or near Sacramento Peak would not appeal to the Evans family, the Evanses must have had a change of heart.

Jack Evans left HAO and Boulder eventually and assumed the directorship of the Sacramento Peak Observatory in 1952. He remained there for many years.

84. McCray, *Giant Telescopes*, 36. Lankford, *American Astronomy*, discusses details of the funding of U.S. astronomy prior to WWII, and the role of powerful observatory directors (chapter 7).

85. Harlow Shapley to Walter Roberts, 25 July 1947, 1.

86. DeVorkin, "Who Speaks for Astronomy?" gives a full discussion of these tensions in this and the following decade. See Menzel's letter to Walter Roberts, for example, expressing a desire to become involved in "confidential" work at Los Alamos, 18 July 1949, fol. Correspondence-Confidential-Menzel, box 35, WOR/UCA.

87. Kevles, *The Physicists*, 355.

88. Alvin G. McNish to A. G. Hill, 4 January 1949, fol. Personal, January 1, 1949 to March 31, 1949, Alan Shapley Collection, Library of Congress, Washington, D.C. (hereafter AHC/LOC). Because of Shapley's extensive involvement in sun–earth science during his long career, this collection represents a treasure trove of documentation relating to space sciences from the late 1940s to the early 1980s. Unfortunately, the LOC has not yet processed this 80+ box collection and is unlikely to do so in the foreseeable future. For a discussion of the creation and demise of the Research and Development Board, see Daniel J. Kevles, "K1S2: Korea, Science, and the State," in *Big Science: The Growth of Large-Scale Research*, eds. Peter Galison and Bruce Hevly (Stanford: Stanford University Press, 1992), 312–333.

89. Kevles, "K1S2," 316.

90. The literature related to post-WWII science policy and the creation of the NSF is extensive, often discussing the episode from different perspectives. Vannevar Bush's *Science—The Endless Frontier*, published in 1945 after President Franklin Roosevelt's request to Bush to think about postwar science, set the stage for the discussions relating to science policy in this era. In *The Physicists*, chapters 11–12, Kevles presents a general overview and discusses the various debates of this period. Daniel S. Greenberg, *The Politics of Pure Science* (Chicago: University of Chicago Press, 1999), analyzes the politics of this era (chapter 6), and concludes that as U.S. science achieved more affluence, it became more rudderless. J. Merton England, *A Patron for Pure Science: The National Science Foundation's Formative Years, 1945–57* (Washington, D.C.: National Science Foundation, 1982), discusses in detail the NSF's early years, including Shapley's important role in the various controversies relating to the NSF's creation. Toby A. Appel, *Shaping Biology: The National Science Foundation and American Biological Research, 1945–1975* (Baltimore: Johns Hopkins University Press, 2000), explains how the NSF moved from simply funding existing research to shaping U.S. scientific disciplines, using their efforts in biological research as an exemplar. Harvey M. Sapolsky, *Science and the Navy: The History of the Office of Naval Research* (Princeton: Princeton University Press, 1990) presents in detail the formation and guiding philosophy of this important institution for U.S. science policy in these years, for the Navy had a long tradition of technological and scientific

innovation. Kevles in "K1S2," 312–333, discusses the postwar development of national security–related science policy and how the Korean War affected changes in this policy-making apparatus.

91. Reported in Wang, *American Science*, 32.

92. Kevles, *The Physicists*, 364.

93. Shapley, *Through Rugged Ways*, 146. Shapley wrote that this was the "one time" he and Bush agreed.

94. Wang, *American Science*, demonstrates in details the tribulations many scientists faced in this difficult period.

95. Hufbauer, *Exploring the Sun*, 132.

96. Roberts, interview by Jackson.

97. DeVorkin, "Donald Menzel," 115; McCray, *Giant Telescopes*, 36–37.

98. DeVorkin, "Who Speaks for Astronomy?" 61.

99. DeVorkin, "Who Speaks for Astronomy?" 61.

100. Shapley to Roberts, 25 July 1947, 1.

101. Kevles, *The Physicists*, 162.

102. England, *A Patron for Pure Science*, 40. This work gives a complete discussion of Shapley's important role in the NSF controversy over the exact form such a foundation might have. Shapley's support of the NSF legislation also put him under increased suspicion by the FBI's J. Edgar Hoover (see Wang, *American Science*, 80).

103. DeVorkin, "Donald Menzel," 115.

104. Shapley to Roberts, 25 July 1947, enclosure.

105. Robert Stearns to Eugene Millikin, 11 August 1947, CA/Pres Off I-100-5, Presidents' Office Files, CUA.

106. DeVorkin, *Science with a Vengeance*, 232.

107. This attitude is obvious in Roberts's response to Menzel's suggestion that HAO staff consider supporting Los Alamos scientists, 19 July 1949, fol. Correspondence-Confidential-Menzel, box 35, WOR/CUA.

108. See Forman, "Behind Quantum Electronics," for an example of a discipline where military oversight was more problematic for those involved.

109. Doel, *Solar System Astronomy*, 193.

110. Roberts to Shapley, 13 August 1947, 1.

111. For Tuve's views on the expansion of the state-sponsorship of science, see Allan A. Needell, *Science, Cold War and the American State: Lloyd V. Berkner and the Balance of Professional Ideals* (Amsterdam: Harwood Academic Publishers, 2000), 269; and Forman, "Behind Quantum Electronics," 218. Needell's work also discusses in detail Berkner's important role in creating many of the organizations that formulated U.S. science policy in this era.

112. Roberts, interview by Jackson.

113. DeVorkin, "Who Speaks for Astronomy?" 71.

114. DeVorkin, "Donald Menzel"; Hufbauer, *Exploring the Sun*, 129–130.

115. Doel, *Solar System Astronomy*, 68.

116. DeVorkin, "Who Speaks for Astronomy?" discusses the rifts of this period in some detail.

117. Doel, *Solar System Astronomy*, 72.

Notes to Chapter 4

1. Ronald E. Doel, "The Earth Sciences and Geophysics," in *Companion to Science in the Twentieth Century*, eds. John Krige and Dominque Pestre (London: Routledge, 2003), 402.

2. Forman, "Behind Quantum Electronic," 211–212.

3. Christophe Lécuyer, *Making Silicon Valley: Innovation and the Growth of High Tech, 1930–1970* (Cambridge: MIT Press, 2007), 5.

4. For a discussion of O'Day's work and his use of V-2 rockets for space science research after WWII, see DeVorkin, *Science with a Vengeance.*

5. See DeVorkin, *Science with a Vengeance*, 221–224; Doel, "The Earth Sciences" and "Constituting the Post-War Earth Science," and Hufbauer, *Exploring the Sun*, chapter 4.

6. William B Pietenpol, *Airborne Coronagraph for Rocket Installation Progress Report #1: April 1, 1948 to July 1, 1948* (Boulder: University of Colorado, 1948). See also report numbers 2–5 for the detailed early development of the UAL.

7. Bartlett and Kraushaar, *A History*, 11–23.

8. Kevles, *The Physicists*, 192.

9. Knight, "Ike Liked the Labs," 2.

10. Kevles, *The Physicists*, 192.

11. O'Mara, *Cities of Knowledge*, 28.

12. Kevles, *The Physicists*, 345.

13. See, for example, Thomas C. Lassman, "Government Science in Postwar America," *Isis* 96, no. 1 (2005): 48. He points out that a fear of nuclear strike on Washington caused Truman in early 1949 to want to move federal facilities. How this fear fit into the CRPL move, as the text points out, is problematic. As for the Eisenhower claim, although Eisenhower did dedicate the new lab, his administration had no part in site selection.

14. See, for example, the discussions of Condon's role in bringing CRPL to Boulder in Lassman, "Government Science" and Wilbert F. Snyder and Charles L. Bragaw, *Achievement in Radio: Seventy Years of Radio Science, Technology, Standards, and Measurement at the National Bureau of Standards*, NBS Special Publication 555 (Boulder, CO: National Bureau of Standards, 1986). The latter is the definitive account of CRPL's history.

15. Edward Condon to Emilie Condon, 8 July 1949, Alan Shapley Files, box 4, fol. 1, Alan H. Shapley Collection/CUA. Clearly postwar strategic considerations such as the fear of another Pearl Harbor–like attack influenced the nation's landscape for science and technology. See O'Mara, *Cities of Knowledge*, 29–45, and Jennifer S.

Light, *From Warfare to Welfare: Defense Intellectuals and Urban Problems in Cold War America* (Baltimore: Johns Hopkins University Press, 2003), chapter 1, for a discussion of postwar dispersion policy and its effects on city panning and suburbanization.

16. The University of Colorado, HAO, and Denver University sponsored the conference.

17. Edward Condon to Emilie Condon, 8 July 1949, 5. Unfortunately for Condon's desires, this was not to be. Condon left the NBS in 1952, before he had full opportunity to take advantage of the CRPL's new location. However, he did eventually return to Boulder in the early 1960s as a professor in the physics department and as a fellow at the Joint Institute for Laboratory Astrophysics.

18. See, for example, Lassman, "Government Science." See also Snyder and Bragaw, *Achievement in Radio*, 708nn16–17.

19. Forman, DeVorkin, and Doel barely mention the important role politicians played in U.S. science in this period. Forman, "Behind Quantum Electronics"; DeVorkin, *Science with a Vengeance*, and "Who Speaks for Astronomy?"; Doel, "Constituting the Postwar Earth Sciences," and "Earth Science and Geophysics."

20. Edwin Johnson to Francis Reich, February 14, 1967, National Bureau of Standards, Special Manuscripts (hereafter NBS file), CUA, 5.

21. Snyder and Bragaw, *Achievement in Radio*, 715n36.

22. For Johnson's role in Rocky Flats, see Len Ackland, *Making a Real Killing: Rocky Flats and the Nuclear West* (Albuquerque: University of New Mexico Press, 1999). Although this site is only about 15 miles from Boulder, its story is more about regional development. Even though built in the mid-1950s, and contemporaneous to developments in Boulder, the story of Rocky Flats relates to Boulder's development as a center for space and atmospheric science tangentially, with Senator Johnson as the linking factor. In 1954, an NBS cryogenics lab came to Boulder, possibly to support Rocky Flats, but this facility did not play a large role in the development of Boulder science.

23. Despite serving on these important committees and being cosponsor of the controversial May–Johnson bill giving all control of atomic matters to the military, Johnson gained his greatest notoriety in 1954 after he denounced actress Ingrid Bergman for leaving her husband to run off with Italian director Roberto Rossellini. Johnson also was on the committee that censured his fellow senator, Joe McCarthy. Colorado State, "The Governor Edwin Carl Johnson Collection at the Colorado State Archives," http://www.colorado.gov/dpa/doit/archives/govs/johnson.html (site discontinued).

24. Charles Sawyer to Wallace White, 26 May 1948, NBS file/CUA, 1.

25. Johnson to Reich, 14 February 1949, NBS file/CUA, 6.

26. Edward Cooper, 6 December 1966, memo, NBS file/CUA, 1.

27. Johnson to Reich, 14 February 1949, NBS file/CUA, 7.

28. Sawyer to Johnson, 24 February 1949, attachment, NBS file/CUA, 4.

29. Sawyer to Johnson, 2 June 1949, NBS file/CUA, 1.

30. Cooper, 6 December 1966, memo, NBS file/CUA, 2.

31. Edward Condon to Emilie Condon, 8 July 1949, 2, 4.

32. Edward Condon to Emilie Condon, 8 July 1949, 1–2.

33. Roberts, interview by Gassaway.

34. Edward Condon to Emilie Condon, 8 July 1949, 4.

35. Knight, "Ike Liked the Labs," 2.

36. Rosenberg, *Boulder, Colorado*, 63.

37. In-depth works on the development of regional centers for science and technology such as O'Mara, *Cities of Knowledge*, and Leslie, *The Cold War*, present various scenarios of regional development, but none are analogous to Boulder.

38. Markusen et al., *The Rise of the Gunbelt*, 178–180.

39. Roger Launius, interview by Joseph Bassi, February 3, 2009.

40. Markusen et al., *The Rise of the Gunbelt*, 181–183; Johnson Space Center, "JSC Origins," NASA, last modified May 17, 2008, http://www.nasa.gov/centers/johnson/about/history/jsc40/jsc40_pg3.html.

41. Stuart W. Leslie, "'A Different Kind of Beauty:' Scientific and Architectural Style in I. M. Pei's MESA Laboratory and Louis Kahn's Salk Institute," *Historical Studies in the Natural Science* 38, no. 2 (2008): 205.

42. Outside funding for University of Colorado research totaled $1 million in 1949. See Davis, *Glory Colorado*, 540.

43. "$1,500,000 Electronics Research Lab of Bureau of Standards May Be Moved Here," *Boulder Daily Camera* (hereafter *BDC*), August 22, 1949, NBS files, Carnegie Library, Boulder, CO.

44. Walter Roberts to John Allardice, 24 August 1949, Misc Confidential Notebook, 1949–1957, WOR/UCARA.

45. Alan Shapley to Walter Roberts, 25 August 1949, National Bureau of Standards file, box 10, WOR/CUA.

46. Roberts to Shapley, 8 September 1949, WOR.

47. Site Selection Board, 12 December 1949, report, National Bureau of Standards, NBS files, Carnegie Library, Boulder, CO.

48. Roberts to Shapley, 7 October 1949, WOR/CUA.

49. Roberts to Shapley, 13 October 1949, WOR/CUA.

50. Walter Roberts, telephone conversation with Donald Menzel, 11 October 1949, memo, box 35, fol. Correspondence-Confidential-Menzel, WOR/CUA.

51. Roberts, telephone conversation with Menzel.

52. Roberts, telephone conversation with Menzel.

53. "Boulder Has Good Chance of Getting Big Laboratory," *BDC*, October 25, 1949.

54. "Boulder Pushing Fight for Big Lab," *BDC*, November 2, 1949.

55. "Site for Bureau of Standards Laboratory Offered to Meet Any Competition of Other Cities," *BDC*, November 2, 1949.

56. "Mayor Newton Is Supporting Boulder for US Laboratory," *BDC*, November 17, 1949.

57. Site Selection Board, 10.

58. Site Selection Board, 14.

59. Knight, "Ike Liked the Labs," 3.

60. "Boulder Expected to Be Picked as Site of 44,500,000 Lab," *BDC*, December 14, 1949.

61. "Campaign Opens April 10 for U.S. Standards Bureau," *BDC*, February 27, 1950.

62. "Yeager to Head $70,000 Campaign to Purchase Site for U.S. Laboratory," *BDC*, March 3, 1950.

63. "Top U.S. Official Visits Boulder on Radio Lab Project," *BDC*, March 2, 1950.

64. "City Council Approves C. of C. Work in Bringing Laboratory to Boulder," *BDC*, March 22, 1950.

65. "Ten Campaign Captains Named," *BDC*, March 18, 1950.

66. "Bureau of Standards Laboratory Brings City Publicity That Is Not 'For Sale,'" *BDC*, March 27, 1950.

67. "Trades Council Makes Donation to Lab Site Fund," *BDC*, March 18, 1950.

68. "Boulder to Benefit from U.S. Bureau of Standards Lab," *BDC*, March 23, 1950.

69. "Acre of Lab Site Is Purchased by Camera Job Staff," *BDC*, April 6, 1950.

70. "Donations to Campaign for Bureau of Standards Laboratory Tax Free," *BDC*, March 17, 1950.

71. Ad in *BDC*, April 5, 1950.

72. "Prosperity Insurance Policy," NBS, Boulder Historical Society 328, box 169, fol. 11. How the Chamber arrived at these numbers remains unknown.

73. "Lab Site Fund Is Far over Top with $82,404.65," *BDC*, April 20, 1950.

74. "Radio Lab 'Only Beginning for Boulder'—Sen. Johnson," *BDC*, April 20, 1950.

75. "Plaque Honoring Donors to Bureau of Standards," *BDC*, August 25, 1954.

76. "Ike Hails Vast Importance of Boulder Laboratories," *BDC*, September 14, 1954. The NBS and the Atomic Energy Commission had built a Cryogenic Engineering Laboratory on the same site as the new CRPL two years earlier, but this facility was not related to space or atmospheric science in any way. The facility did, however, contribute to the city's growing reputation as a place for modern science and technology, a reputation that fuelled its development in the mid- to late 1960s.

77. Walter A. McDougall, *The Heavens and the Earth: A Political History of the Space Age* (New York: Basic Books, 1985), 387.

78. Vannevar Bush, *Science—The Endless Frontier* (Washington, D.C.: National Science Foundation, 1990), 4.

79. Stephen E. Ambrose, *Eisenhower: Soldier and President* (New York: Touchstone, 1990), 292.

80. Peter Novick, *That Noble Dream: The "Objectivity Question" and the American Historical Profession* (Cambridge: Cambridge University Press, 1988), 88.

81. For a discussion on the importance of these types of associations, sometimes

called "horizontally integrated networks," in the production of scientific and techni-cal knowledge, see Kargon et al., "Far beyond Big Science," 335.

82. Lécuyer, *Making Silicon Valley*, 1.

83. Lécuyer, *Making Silicon Valley*, 1–2.

Notes to Chapter 5

1. *Physics Today* 3 (1950): 35, quoted in Kevles, *The Physicists*, 367.

2. Lankford, *American Astronomy*, 206.

3. D. Roberts, *On the Ridge*, 31.

4. Roberts, interview by McFarland, 2.

5. Walter Roberts to Harlow Shapley, 5 December 1949, attachment, fol. Harlow Shapley Confidential, 1949–1950, box 14, WOR/CUA.

6. Roberts, interview by DeVorkin, 82.

7. It appears that at this time, Roberts did not realize that the "red tape" issue was emblematic of all work sponsored by the government, irrespective of the exact funding source.

8. See Foreman, "Behind Quantum Electronics," and Doel, "Constituting the Postwar Earth Sciences."

9. Roberts, interview by DeVorkin, 82.

10. "Report of the High Altitude Observatory," 6 July 1954, report, reference file, HAO Collection, UCARA, 5.

11. Lockyer, "Progress in Astronomy," 145.

12. Clark, *The Sun Kings*. See chapter 2 for a full discussion of William Herschel's thinking on possible sun–weather relations.

13. Clark, *The Sun Kings*, 68–69.

14. Katharine Anderson, "The Weather Prophets: Science and Reputation in Vic-torian Meteorology," *History of Science* 27 (1999): 203–204.

15. David DeVorkin, "Defending a Dream: Charles Greeley Abbot's Years at the Smithsonian," *Journal for the History of Astronomy* 21 (1990): 121.

16. Thomas Bogdan, "Donald Menzel," 161–162. Jack Eddy, an astronomer at HAO and a long-time student of possible sun–weather connections referred to Menzel's approach as a "snake-oil promise." See "HAO-50th Anniversary Talk: A Tribute to Walter Orr Roberts," 30 October 1990, HAO 50th Anniversary Binder, UCARA, 3.

17. Roberts, interview by DeVorkin, 110.

18. Theodore F. Sterne, Karl Frederick Guthe, and Walter Roberts, "On Possible Changes in the Solar 'Constant,'" *Proceedings of the National Academy of Sciences of the United States* 26, no. 6 (1940), 399–406.

19. Roberts pointed out the need for a new coronagraph site with better air qual-ity farther from the mine as early as 1943 in his doctoral dissertation. *Preliminary Studies of the Solar Corona and Prominences with the Harvard Coronagraph* (PhD diss., Harvard University, 1943), 91.

20. *The High Altitude Observatory of Harvard University and University of Colorado: An Explanation of its Origins, Scientific Objectives, and Future Needs*, January 1952, fol. Colorado, University of, 1952, box 19, WOR/CUA, 2.

21. Walter Roberts to Richard G. Croft, 15 January 1953, fol. Great Northern paper, 1953, box 29, WOR/CUA.

22. Roberts, interview by DeVorkin, 74.

23. Roberts, interview by DeVorkin, 75.

24. Kevles, *The Physicists*, 370.

25. McCray, *Giant Telescopes*, 34–36.

26. Memoranda, June 16, 1953; July 9, 1953, William Andrews to HAO Trustees, Folder 11, Admin: Memoranda-Trustees, HAO, 1953–1960, John L. J. Hart Papers, UCARA; Walter Roberts to Thornton C. Fry, March 25, 1953, Folder: Bell Labs, box 5, WOR/CUA, enclosure.

27. Robert Stearns to L. M. Pexton, 28 April 1953, fol. Colorado, University of, 1953, box 19, WOR/CUA.

28. Robert Low to Robert Stearns, 17 April 1953, fol. Colorado, University of, 1953, box 19, WOR/CUA.

29. Frank Kemp to Robert Stearns, 14 April 1949, fol. Colorado, University of, 1949, box 19, WOR/CUA.

30. James L. Breese to Walter Roberts, 29 April 1952, enclosure, fol. Breese, James L. 52–53, WOR/CUA, 2.

31. Breese to Roberts, 29 April 1952.

32. Walter Roberts to James L. Breese, 30 April 1952, fol. Breese, James L. 52–53, WOR/CUA.

33. James L. Breese to William Andrews, 1 June 1953, fol. Breese, James L. 52–53, WOR/CUA.

34. James L. Breese to William Andrews, 7 July 1953, fol. Breese, James L. 52–53, WOR/CUA.

35. Robert Low to Walter Roberts, 24 September 1953, fol. Breese, James L. 52–53, WOR/CUA.

36. Walter Roberts, "Big Business and Research," *Science*, 121 (1955): 473, doi: 10.1126/science.121.3144.473.

37. Roberts, "Big Business and Research," 473.

38. Wang, *American Science*.

39. The literature on Oppenheimer and his trial is extensive. For a recent analysis of his security difficulties, see Kai Bird and Martin J. Sherwin, *American Prometheus: The Triumph and Tragedy of J. Robert Oppenheimer* (New York: Alfred A. Knopf, 2005), chapters 33–37.

40. Kevles, *The Physicists*, 367.

41. Kevles, *The Physicists*, 379.

42. Roberts, interview by DeVorkin, 86.

43. Wang, *American Scientists*, 129.

44. Ronald E. Doel, "Redefining a Mission: The Smithsonian Astrophysical Observatory on the Move," *Journal for the History of Astronomy* 21 (1990): 141. Bowen's concern arose during a period when the Smithsonian considered moving its observatory to Colorado and under Roberts's direction, an episode discussed in detail subsequently.

45. Roberts, for example, was a member of the MITRE board in the 1960s, a position that required a security clearance.

46. Wang, *American Science*, 86.

47. Doel, "Redefining a Mission," 141. Doel writes that the Air Force denied Roberts clearance to Sac Peak. More precisely, the evidence indicates the issue was access to Holloman Air Force Base, which Roberts needed to travel to en route to Sacramento Peak. Roberts did not do classified work there.

48. Walter Roberts to Alan Shapley, 31 December 1950, attachment dated October 19, 1950, fol. 1, box 4, AHC/CUA.

49. Walter Roberts FBI File, UCARA.

50. Wang, *American Science*, 126.

51. Roberts, interview by DeVorkin, 76.

52. Roberts, interview by DeVorkin, 78.

53. Walter Roberts to Donald Menzel, 14 April 1950, fol. Correspondence-Confidential-Menzel, box 35, WOR/CUA.

54. Walter Roberts to Robert Stearns, 14 July 1950, fol. 7, DHM-WOR Correspondence, June 14, 1950–December 22, 1950, box 1, DHM/DU.

55. Harlow Shapley to Walter Roberts, 20 September 1950, Harlow Shapley Confidential folder, 1949–1950, box 14, WOR/CUA.

56. Walter Roberts to Harlow Shapley, 20 September 1950, fol. Harlow Shapley Confidential, 1949–1950, box 14, WOR/CUA.

57. Walter Roberts to Alan Shapley, 31 December 1950, fol. 1, box 4, AHC/CUA.

58. Roberts to Shapley, 31 December 1950.

59. "Hearing," December 4, 1950, transcript of Roberts's hearing, Munitions Board, Industrial Employment Review Board, UCARA.

60. Roberts, interview by DeVorkin, 88. Roberts also said that he often had received calls from the other Roberts's girlfriends, thinking at the time these incidents were pranks of some sort.

61. Roberts, interview by DeVorkin, 78.

62. Roberts's hearing transcript contains most of the details of this difficult episode. His interview with DeVorkin also relates some of his feelings about the period. Roberts, interview by DeVorkin, 78.

63. DeVorkin, "Donald Menzel."

64. Donald Menzel to Walter Roberts, 4 July 1950, fol. 7, DHM-WOR Correspondence, June 14, 1950–December 22, 1950, box 1, DHM/DU.

65. "Hearing," 53. The typist used capitals for the word in the transcript. Menzel's "Dear Tovarich" letter to Stalin in July 1938 written on behalf of Soviet astronomers,

if known to Roberts's or his own clearance review board, probably did not help his cause. Donald Menzel to Joseph Stalin, July 1938, fol. 2, box 1935–1952, DHM/DU.

66. Donald Menzel to Robert Stearns, 6 July 1950, fol. 7, DHM-WOR Correspondence, June 14, 1950–December 22, 1950, box 1, DHM/DU. In light of Roberts's subsequent enthusiasm and relatively successful fundraising, Menzel's claims seem somewhat problematic, perhaps reflecting his own difficulties in this regard.

67. Roberts to Menzel, 14 April 1950.

68. Walter Roberts to Harlow Shapley, 23 June 1950, fol. Harlow Shapley Confidential, 1949–1950, box 14, WOR/CUA, 1.

69. Roberts to Shapley, 23 June 1950.

70. Menzel to Stearns, 6 July 1950.

71. Roberts, interview by DeVorkin, 83.

72. Harlow Shapley to Walter Roberts, 30 June 1950, fol. Harlow Shapley Confidential, 1949–1950, box 14, WOR/CUA.

73. Roberts, interview by DeVorkin, 83.

74. Hallgren, *The University Corporation*, 61.

75. Walter Roberts to Donald Menzel, 18 April 1951, fol. Correspondence Confidential Menzel, box 35, WOR/CUA, 5. Roberts ended this long and somewhat somber letter with a cheery "We all look forward to seeing you in a few weeks. There is certainly lots of material to cover."

76. See Walter Roberts, 27 February 1952, memo, "Misc. Confidential 1949–1957" notebook, UCARA.

77. Walter Roberts to Donald Menzel, 25 February 1953, fol. Correspondence Confidential Menzel, box 35, WOR/CUA, 1.

78. Roberts to Menzel, 25 February 1953, 1.

79. Donald Menzel to Walter Roberts, 16 March 1953, fol. Correspondence Confidential Menzel, box 35, WOR/CUA, 1.

80. Donald Menzel to Walter Roberts, 13 April 1953, box 17, WOR/CUA.

81. Robert J. Low, 17 September 1953, memo, fol. Menzel-C.S.O., box W. O. Roberts 1940–57, M-N, WOR/CUA. This six-page memo not only describes Higman's observations of the Menzel–Roberts relationship, but also lays out a general strategy that appears, in retrospect, to have worked in bringing an end to the Harvard–Colorado relationship a year later.

82. Low, 17 September 1953, memo.

83. Roberts, interview by Jackson. Roberts in this interchange claimed he was always a "Harvard man" at heart. As a result, he argues that he did not really want the breakup and thought Harvard had made a great mistake in deciding to terminate the joint program. To the extent that this reflects his thinking in 1954, it also is clear from the record that any continuation could only have existed on the Boulder contingent's terms.

84. Darley, in his short three-year tenure as university president, became a strong supporter of Roberts's efforts to bring more sun–earth science to Boulder and the university. In 1955 he joined with Roberts in an unsuccessful effort to bring the

relocating Smithsonian Astrophysical Observatory to Boulder. As Doel describes, a major reason for the SAO decision to not move to Boulder centered precisely on the lack of astrophysical research and training at the university. Doel, "Redefining a Mission," 142–143. This failure resulted in Darley strongly supporting Roberts's effort to create a Department of Astro-Geophysics at the university in 1956.

85. Walter Roberts, 1 March 1954, memo, fol. Colorado, University of, 1954, box 20, WOR/CUA, 5.

86. McGeorge Bundy to Ward Darley, 29 March 1954, fol. I-79-3, President's Office Files, CUA (hereafter PF/CUA).

87. Thomas B. Knowles to McGeorge Bundy, 30 April 1954, fol. I-79-3, PF/CUA, 2.

88. McGeorge Bundy to Ward Darley, 6 August 1954, notebook, John L. J. Hart Papers, V.III, UCARA.

89. McGeorge Bundy to Roberts Stearns, 12 May 1954, fol. I-79-3, PF/CUA.

90. Thomas B. Knowles to Ward Darley, 4 May 1954, fol. I-79-3, PF/CUA.

91. Roberts, interview by Jackson.

92. Bundy to Darley, 6 August 1954.

93. Walter Roberts to Hugh Odishaw, 4 March 1955, fol. I.G.Y., box W. O. Roberts 1940–57, h-k, WOR/CUA.

94. Donald Menzel to Walter Roberts, 14 October 1954, fol. Solar Associates files, AHC/LOC.

95. Knowles to Darley, 4 May 1954.

96. George E. Webb, *Science in the American Southwest: A Topical History* (Tucson: University of Arizona Press, 2002), xix.

97. Even though Sacramento Peak was funded by, and from 1952 operated by, the Air Force, there were always close ties between HAO in Boulder and its former "colony" in New Mexico. The story of Sacramento Peak, in turns, also forms an important aspect of U.S. science's move to the Southwest in the postwar era.

98. Roberts, interview by DeVorkin, 85.

99. Walter Roberts to Ward Darley, 21 September 1954, fol. Colorado, University of, 1954, box 20, WOR/CUA.

100. Hallgren, *The University Corporation*, 65.

101. Roberts, interview by DeVorkin, 116.

102. Walter Roberts to William A. M. Burden, 26 September 1955, fol. Correspondence "B", 1955, box 1, WOR/CUA.

103. Walter Roberts to Arthur H. Bunker, 26 September 1955, fol. Correspondence "B", 1955, box 1, WOR/CUA.

104. Walter Roberts to Frederick W. Conant, 11 May 1956, fol. Correspondence "C", 1956, box 3, WOR/CUA. In the 1930s, Julius Bartels proposed "M regions" as non-sunspot related sources of streams of solar particles that cause geomagnetic storms on earth. Bartels's insight proved correct, and solar physicists today refer to the speculated M regions as "coronal holes." These structures appear most noticeably in x-ray images of the solar surface.

105. Walter Roberts to Frederick W. Conant, 11 May 1956.

106. Roberts, interview by DeVorkin, 119. To this day, the idea that solar activity and its geophysical effects may affect the earth's weather remains controversial—hence Roberts's use of the term witchcraft in describing the views of scientists who thought such relationships unlikely. Proponents of a sun–weather connection, including Roberts, have never offered a convincing direct physical link between the processes residing in the thin upper atmosphere and the much denser lower atmosphere. Even Roberts admitted much work in this field was "poorly done" and based on "sloppy" statistics. Barbara B. Poppe and Kristen P. Jordan, *Sentinels of the Sun: Forecasting Space Weather* (Boulder: Johnson Books, 2006), 41.

107. Douglas V. Hoyt and Kenneth H. Schatten, *The Role of the Sun in Climate Change* (New York: Oxford University Press, 1997), 114. Noted atmospheric scientist R. A. Craig wrote this in 1965 describing the views of many meteorologists.

108. Roberts, interview by DeVorkin, 114.

109. Donald Menzel, *Our Sun* (Philadelphia: The Blakiston Company, 1949), 311. In this solar physics work, Menzel included an entire section on both sun–weather research and another on "other solar-terrestrial relations," or the sun–earth connection, in other words. These sections were in a chapter not-so-subtly titled "The Sun—And You!" (chapter 12, 286–315).

110. Walter Roberts to Leonard Carmichael, 5 February 1955, fol. Smithsonian Institution 1954, box 13, WOR/CUA. It was Roberts's desire to complete this interdisciplinary and controversial research that made him a strong candidate to lead a relocated Smithsonian Astrophysical Observatory in the first place. See Doel, "Redefining a Mission," 139–140. His interest in sun–weather connection research put Roberts in the footsteps of the Smithsonian's C. G. Abbot's efforts on the effect of solar luminosity variations and earth weather. However, the dearth of advanced scientific training in Boulder proved fatal to Carmichael's (and Roberts's) desires.

111. Walter Roberts to Francis Reichelderfer, 28 April 1955, fol. R 1955, box W. O. Roberts 1940–1957 r-s, WOR/CUA.

112. Doel, *Solar System Science*, 198. Astrogeophysics is of course a somewhat obtuse, albeit shorter, way of saying sun–earth connection science. The origin of this term is murky, and sometimes (in the author's experience in this field of research) Sydney Chapman received credit among scientists for creating this shorthand for the scientific research related to the sun–earth connection in the early 1950s. To obfuscate the matter further, some scientists used the term to describe the interaction of the sun with any planet in the solar system—in effect rendering it a substitute for solar system studies.

113. Doel, *Solar System Astronomy*, 26–27, 199.

114. Donald Menzel, "Astronomy 220 Syllabus," 1954, fol. Donald H. Menzel 1954, box 17, WOR/CUA.

115. Doel, *Solar System Astronomy*, 205.

116. Doel, *Solar System Astronomy*, 197.

117. It is important to note that this was an era of increasing fragmentation of astronomy into subdisciplines based on new instrumental approaches and increasing specializations (Doel, *Solar System Astronomy*, 197). Some traditional astronomers lamented the "de-astronomization" of astronomy (McCray, *Giant Telescopes*, 30).

118. Doel, *Solar System Astronomy*, 205.

119. Doel, "Redefining a Mission," 143. Doel argued that the lack of a good astronomy department at the University of Colorado and distance from the East Coast, not Roberts's security problem in 1950 or his political activism, resulted in the Smithsonian moving the Observatory to Harvard. This conforms to Roberts's view as to why Boulder did not get the SAO.

120. "Report of the Graduate School Committee on the Proposed Astro-Geophysics Program," 4 October 1956, fol. Colorado, University of, Proposed Astro-Geophysics Program 1956, box 20, WOR/CUA.

121. Roberts, interview by DeVorkin, 122.

122. Sullivan, *Assault on the Unknown*.

Notes to Chapter 6

1. Sullivan, *Assault on the Unknown*, 4.

2. Fae L. Korsmo, "The Genesis of the International Geophysical Year," *Physics Today*, July 2007, 38.

3. James H. Capshew and Karen A. Rader, "Big Science: Price to the Present," *Osiris, Second Series* 7 (1992): 23. In this article they discuss big science as "Pathology," "Scientific Phenomenon," "Instrument," "Industrial Production," "Ethical Problem," "Politics," Institution," "Culture," and "Form of Life." However, as Andrew Butrica wrote, "Defining Big Science is the intellectual equivalent of trying to nail Jell-O to the wall." Andrew J. Butrica, *To See the Unseen: A History of Planetary Radio Astronomy* (Washington, D.C.: NASA History Office, 1996), ix. This work's introduction presents an interesting discussion of the idea of big science, but within the context of planetary radio astronomy. Alvin M. Weinberg, director of Oak Ridge National Laboratory, helped to make the term "Big Science" popular in the early 1960s as he lamented what he considered its deleterious effects on modern research. Kevles, *The Physicists*, 392, 417. The literature on big science and its implications is extensive; see, for example, Peter Galison and Bruce Hevly, eds., *Big Science: The Growth of Large-Scale Research* (Stanford: Stanford University Press, 1992). This work consists of chapters on many disparate, but important, facets of modern U.S. big science. For the effect of this postwar era of big science on the nation's universities, see Geiger, "Science, Universities, and National Defense," 26–48. McCray in *Giant Telescopes* discusses how the drift toward big science has, and continues to, affect the practice of modern astronomy and almost produce unease among some of its practitioners (4–9).

4. Needell, *Science, Cold War*, 319.

5. Capshew and Rader, "Big Science," 19.

6. Abigail Foerstner, *James Van Allen: The First Eight Billion Miles* (Iowa City: University of Iowa, 2007), 123. Also see Needell, *Science, Cold War*.

7. See DeVorkin's *Science with a Vengeance* for a complete discussion of the military use of V-2 for postwar high-altitude scientific research. Korsmo, "The Genesis," 38–43, presents an analysis of the scientific and political context that led to this dinner and the subsequent IGY planning that arose from it. For a brief discussion of the range of new scientific tools then available that stimulated thinking about IGY, see Sullivan, *Assault on the Unknown*, 22–23.

8. Clark, *The Sun Kings*, 52–53; Peter J. Bowler, *The Environmental Sciences* (New York: W. W. Norton, 1993), 209. Baron von Humboldt advocated an interdisciplinary (or holistic) approach to investigating the natural world scientifically. He was also a pioneer of scientific fieldwork as he undertook his "global physics." David N. Livingstone, *Putting Science in Its Place: Geographies of Scientific Knowledge* (Chicago: University of Chicago Press, 2003), 40, 4. Sun–earth connection research, such as the relationship between the sun and earth's magnetic variations, was therefore a manifestation of what scholars call "Humboldtian science." Bowler, *The Environmental Sciences*, 205. Roberts, with his broad interests in the sun–earth connection in its manifest forms, certainly fits the description of a "Humboldtian scientist."

9. Fae L. Korsmo and Michael P. Sfraga, "From Interwar to Cold War: Selling Field Science in the United States, 1920s through 1950s," *Earth Science History* 22, no. 1 (2003): 64.

10. Korsmo and Sfraga, "From Interwar to Cold War," 65.

11. This is the same James Webb that later led NASA in the Apollo era, taking over the organization in the Kennedy administration.

12. Needell, Science, *Cold War*, 146.

13. Lloyd V. Berkner, "International Scientific Action: The International Geophysical Year 1957–58," *Science* 119 (1954): 575, doi:10.1126/science.119.3096.569.

14. Needell, *Science, Cold War*, 298.

15. Needell, *Science, Cold War*, 302.

16. Berkner, "International Scientific Action," 572.

17. Dorothy Schaffter, *The National Science Foundation* (New York: Praeger, 1969), 147.

18. United States National Committee of the International Geophysical Year 1957–58, 26 March 26 1953, minutes, fol.: US-IGY 1st Meeting, unprocessed, Alan H. Shapley Collection, Library of Congress, Washington, D.C. (hereafter USNC).

19. USNC, 16.

20. Homer E. Newell, "Beyond the Atmosphere: Early Years of Space Science," NASA SP-211 (Washington, D.C.: U.S. Government Printing Office, 1980), 51.

21. Berkner, "International Scientific Action," 12.

22. Needell, *Science, Cold War*, 304. For a discussion of Tuve's attitude, see p. 308.

23. Jacob Darwin Hamblin, *Oceanography and International Cooperation during*

the *Early Cold War* (PhD diss., University of California, Santa Barbara, June 2001). Hamblin presents a thorough analysis of the diverse, and often conflicting, views the U.S. oceanographic community held on scientific cooperation on an international scale in this period.

24. Walter Roberts to Ward Darley, 15 January 1956, fol. Colorado, University of, 1956, box 20, WOR/CUA.

25. "Twenty-Fourth Award of the William Bowie Medal," 26 April 1962, transcript of award prepared and presented by Walter Roberts, NBS/CRPL-HAO joint file, WOR/CUA.

26. Joseph Kaplan to Walter Roberts, 10 December 1954, fol. IGY, box 1940–1957 h-k, WOR/CUA.

27. Walter Roberts to Joseph Kaplan, 13 December 1954, fol. IGY, box 1940–57 h-k, WOR/CUA.

28. Crowe et al., *Remembering Walter Roberts*, 98.

29. Needell, *Science, Cold War*, 319.

30. Charles E. Rosenberg, *No Other Gods: On Science and American Social Thought* (Baltimore: Johns Hopkins University Press, 1997), 2225–2239.

31. Needell, *Science, Cold War*, 300.

32. See chapter 3 for the discussion of this episode. Odishaw was NBS director Edward Condon's assistant.

33. Needell, *Science, Cold War*, 309.

34. Stanley Ruttenberg, interview by Helen Coffey and Diane Rabson, July 23, 2007, Tape AMS 218–219, UCARA, Boulder, CO. I am indebted to Diane Rabson for graciously asking questions during the interview that this author proposed.

35. Pembroke Hart, interview by Joseph Bassi, December 14, 2006, Washington, D.C. Hart was an official of the IGY committee.

36. Alan Shapley, n.d., notes, fol. IRE Advisory Committee-1958, unprocessed collection, Alan H. Shapley Collection, Library of Congress, Washington, D.C. (hereafter AHC/LOC).

37. "Proposed Mission of the Central Radio Propagation Laboratory," n.d., fol. CRPL-IGY 1958, unprocessed collection, AHC/CUA.

38. Alan Shapley to Alan Astin, 18 February 1954, fol. IGY WDC Info 1955–1957, unprocessed collection, AHC/CUA, 4.

39. U.S. National Committee for the IGY, "14th Meeting," 16 June 1958, notes, fol. USNC, IGY-1958, unprocessed collection, AHC/LOC. The other panels were Seismology and Gravity, Glaciology, Longitude and Latitude, Rocketry, and Satellites.

40. Walter Roberts to Gordon Allott, 4 February 1956, fol. IGY 1956, box 1940–1957 h-k, WOR/CUA.

41. Obtaining this government funding for an IGY with a broad scientific agenda required extensive coordination and lobbying on the part of Odishaw, Shapley, and others. See Korsmo and Sfraga, "From Interwar to Cold War," for a comprehensive description of these events.

42. Richard T. Hansen to Hugh Odishaw, 19 June 1957, fol. HAO IGY 1957, box 1940–57, h-k, WOR/CUA.

43. Korsmo, "The Genesis," 42; Sullivan, *Assault on the Unknown*, 35.

44. Needell, *Science, Cold War*, 315.

45. Quigg Newton to Gordon Allott, 29 August 1957, fol. HAO IGY 1957, box 1940–57 h-k, WOR/CUA. This is the draft of the letter, but subsequent correspondence indicates the letter did get to the senator.

46. J. Tuzo Wilson, *IGY: The Year of the New Moons* (New York: Alfred A. Knopf, 1967), 122–124. Merle Tuve, Berkner's mentor, invented the ionosonde in the 1920s.

47. Sullivan, *Assault on the Unknown*, 170–172.

48. Sullivan, *Assault on the Unknown*, 216–219.

49. Alan Shapley, "Outline of CRPL Participation in the IGY Program," 2 April 1958, fol. CRPL-IGY 1958, unprocessed collection, AHC/LOC.

50. Alan Shapley, "Summary of Oral Report, World Days and Communications," 16 June 1958, fourteenth meeting, USNC-IGY, fol. USNC-IGY 1958, unprocessed collection, AHC/LOC.

51. Berkner, "International Scientific Action," 575.

52. Alan Shapley to "TNG et al.," 3 September 1954, fol. Miscellaneous, unprocessed collection, AHC/LOC. The original copy appears to remain in this folder, so it may have never been sent to the intended recipient.

53. Shapley to Astin, 18 February 1954.

54. Synder and Bragaw, *Achievement in Radio*, 452–453. This service continued, via the NBS radio station (WWV), until October 1976. The Ft. Belvoir warning location operated until 1968 and then relocated to Boulder. In 1965, the site's name changed to the somewhat more descriptive "Telecommunications and Space Disturbance Center." For ease of transmission, WWV broadcast the warning in Morse code until 1971, and then switched to voice announcements. CRPL staff based these warnings on such indicators as the nature of activity on the sun's surface and its coronal regions, the secular changes in the ionosphere, and the state of the earth's magnetic field in any given time period. In the early 1960s, the ensemble of these services and supporting science fell under the rubric of "space weather."

55. Alan Shapley to A. G. McNish, 17 July 1953, 1953 Chronological File, box 43, AHC/LOC.

56. Alan Shapley to Alan Astin, 19 August 1953, AHC/LOC.

57. Alan Shapley, "CRPL Participation in the International Geophysical Year," December 1955, fol. IGY WDC Info, 1955–1957, AHC/LOC.

58. Synder and Bragaw, *Achievement in Radio*, 455.

59. Walter Roberts, "Summary of Oral Report, Solar Activity," 16 June 1958, fourteenth meeting, USNC-IGY, fol. USNC-IGY 1958, unprocessed collection, AHC/LOC.

60. Sydney Chapman, "Guide to the IGY World Data Centers," n.d., fol. CSAGI to IGY WDC, box 57, unprocessed collection, AHC/LOC, 1.

61. Korsmo, "The Genesis," 42. The implementation of this policy on a worldwide scale proved problematic since, for example, simply getting to an IGY center in the Soviet Union to study data sets proved difficult. Nevertheless, this open exchange of scientific information was IGY's goal.

62. Hufbauer, *Exploring the Sun*, 155.

63. Hufbauer, *Exploring the Sun*, 156.

64. Sullivan, *Assault on the Unknown*, 34–35.

65. The story of the in situ calibration and actual collection of these vast, disparate, and complex data sets collected during the IGY period is well beyond the scope of this book, and perhaps worth its own treatment.

66. Alan Shapley to Francis Brown, 15 May 1956, fol. IGY WDC Info 1955–57, unprocessed collection, AHC/LOC, 2.

67. Shapley to Brown, 15 May 1956, 2.

68. E. H. Vestine to Ward Darley, 5 June 1956, fol. IGY 1956, box 1940–57, h-k, WOR/CUA.

69. WOR to Project Scientific Officer, CRPL, 25 November 1955, fol. HAO-CRPL Joint Program, unprocessed collection, AHC/LOC.

70. Vestine to Darley, 5 June 1956.

71. Alan Shapley to Joseph Kaplan, 9 July 1956, fol. USNC-IGY Vice Chairman 1955–56, box 13, unprocessed collection, AHC/LOC, 2.

72. Shapley certainly must have found himself in a number of awkward situations in this period. He was relatively junior in the NBS system, yet he could write directly to NBS director and USNC member Astin because of Shapley's role as the USNC vice chair.

73. Walter Roberts to E. O. Hurlburt, 4 July 1956, fol. World Data Center, box 45, unprocessed collection, AHC/LOC, 1.

74. Wesley Brittin and Walter Roberts to Ward Darley, 22 June 1956, fol. University of Colorado, 1956, box 20, WOR/CUA.

75. Under the WDC structure, all WDCs had complete sets of data, but only one center was the "chief data center" for a given discipline. The other two WDCs therefore had secondary centers as the backup to the chief data center for that discipline.

76. Brittin and Roberts to Darley, 22 June 1956.

77. Brittin and Roberts to Darley, 22 June 1956.

78. Walter Roberts, 6 September 1956, memo, fol. University of Colorado, 1956, box 20, WOR/CUA.

79 Walter Roberts to Wally Lovelace, 30 November 1956, fol. University of Colorado, 1956, box 20, WOR/CUA, 2.

80. Hugh Odishaw to USNC-IGY Panel Members, 26 December 1956, fol. IGY 1956, box 1940–57 h-k, WOR/CUA.

81. Roberts, interview by DeVorkin, 115.

82. Crowe et al., *Remembering Walter Roberts*, 47.

83. Roberts, interview by Jackson.

84. Kevles, *The Physicists*, 383.

85. Zuoyue Wang, *In Sputnik's Shadow: The President's Science Advisory Committee and Cold War America* (New Brunswick, NJ: Rutgers University Press, 2009), 61.

86. Daniel S. Greenburg, *The Politics of Pure Science* (Chicago: University of Chicago Press, 1999), 273.

87. Newton to Allott, 29 August 1957.

88. "HAO Supervisory Committee Notes, September 5, 1957," 24 September 1957, memo, fol. Administrative, Supervisory Committee Notes, 1957–1960, HAO/UCAR.

89. Newton to Allott, 29 August 1957.

90. Robert Low, "Trustees Memorandum," 17 September 1957, fol. 11: Admin: Memoranda-Trustees, HAO 1953–1960, John L. J. Hart Papers, HAO/UCAR.

91. Walter Roberts to Hugh Odishaw, 28 August 1957, fol. HAO IGY 1957, box 1940–1957 h-k, WOR/CUA. Newton had recently served as the Democratic mayor of nearby Denver.

92 Walter Roberts to J. P. Hagen, 28 August 1957, fol. HAO IGY 1957, box 1940–1957 h-k, WOR/CUA.

93 Newton to Allott, 29 August 1957.

94. Hugh Odishaw to Walter Roberts, 30 August 1957, fol. HAO IGY 1957, box 1940–1957 h-k, WOR/CUA.

95. Hugh Odishaw to Earl Droessler, 30 August 1957, fol. HAO IGY 1957, box 1940–1957 h-k, WOR/CUA.

96. Walter Roberts to Hugh Odishaw, 6 September 1957, fol. HAO IGY 1957, box 1940–1957 h-k, WOR/CUA.

97. Low, "Trustees Memorandum," 3.

98. B. Van Mater to Walter Roberts, 1 October 1957, fol. HAO IGY 1957, box 1940–1957 h-k, WOR/CUA.

99. Crowe et al., *Remembering Walter Roberts*, 47.

100. Roberts, interview by Jackson.

Notes to Chapter 7

1. Roger, L. Geiger, "What Happened after Sputnik? Shaping University Research in the United States," *Minerva* 35 (1997): 353.

2. Davis, *Glory Colorado*, 672.

3. Davis, *Glory Colorado*, 672.

4. Shakespeare, *Julius Caesar*, act 4, scene 3, lines 218–219. Brutus speaks.

5. Quigg Newton, n.d., recollection, fol. 32, box 4, Quigg Newton papers, Denver Public Library, Denver, Colorado (hereafter QN/DPL).

6. Kevles, *The Physicists*, 382.

7. Geiger, "Science, Universities, and National Defense," 43. Roberts never did give up fundraising altogether and found various humanitarian-related projects, such as the Aspen Institute, to lend his fundraising talents. However, his scientifically

related work after 1960 relied almost exclusively on government (primarily NSF) funding.

8. Hufbauer, *Exploring the Sun*, 165.

9. Geiger, "What Happened after Sputnik?" 355.

10. Geiger, "What Happened after Sputnik?" 356.

11. Geiger, "Science, Universities, and National Defense," 43.

12. Lécuyer, *Making Silicon Valley*, 5.

13. Saxenian, *Regional Advantage*, 2. She uses the terms "social networks" to describe the relevant community, but in a casual, as opposed to theoretical, sense.

14. Kevles, *The Physicists*, 386.

15. Greenburg, *The Politics of Pure Science*, 273.

16. McDougall, *The Heavens and the Earth*, 132. This work contains a detailed analysis of the U.S. reaction to *Sputnik* and the perceived Soviet challenge to U.S. technological and cultural superiority presented by its launch.

17. The IGY symbol prominently features a lone satellite orbiting the earth.

18. James Warwick, M. Bretz, and W. O. Roberts to USNC-IGY, 12 November 1957, fol. HAO IGY 1957, box 1940–1957 h-k, WOR/CUA.

19. Walter Roberts to A. C. Bekaert, 1 November 1957, fol. Correspondence: "B" 1957, box 2, WOR/CUA.

20. See Zuoyue Wang, *American Science*, for a complete discussion of overarching U.S. science policy in the 1950s.

21. Meg Greenfield, quoted in Greenberg, *The Politics of Pure Science*, 126.

22. See Greenburg, *The Politics of Pure Science*, 120.

23. For detailed early histories of NSF, see England, *A Patron for Pure Science*; Milton Lomask, *A Minor Miracle: An Informal History of the National Science Foundation* (Washington, D.C.: NSF, 1976); Schaffter, *The National Science Foundation*.

24. See the discussions of this period in Geiger, "Science, Universities, and National Defense" and McDougall, *The Heavens and the Earth*.

25. McDougall, *The Heavens and the Earth*, 149. See also Wang, *American Science*, for an insightful discussion of the response of the Eisenhower administration to *Sputnik*, focusing on the restructuring of U.S. presidential science advising.

26 Schaffter, *The National Science Foundation*, 45.

27. Louis S. Rothschild to Detlev W. Bronk, 27 September 1955, fol. COM 1955–1956, National Academy of Sciences, Committee on Meteorology, National Academy of Sciences, Washington, D.C. (hereafter COM).

28. See Kristine C. Harper, "Meteorology's Struggle for Professional Recognition in the USA (1900–1950)," *Annals of Science* 63 (2006), 179–199, doi:10.1080/00033790600554627.

29. For a discussion of the history of these techniques, see Kristine C. Harper, "Boundaries of Research: Civilian Leadership, Military Funding, and the International Network Surrounding the Development of Numerical Weather Prediction in the US," (PhD diss., Oregon State University, 2003).

30. See COM, "Minutes of Second Meeting" 19 September 1956, fol. COM 1955–1956, COM, and letter from F. W. Reichelderfer to Detlev Bronk, 14 March 1956, fol. COM, 1955–56, COM.

31. Detlev Bronk to Lloyd Berkner, 14 December 1955, fol. COM, 1955–1956, COM.

32. "Importance of New Concepts in Meteorology," 3 April 1956, draft COM report, fol. COM, 1955–1956, COM. See also Rothschild to Bronk, 27 September 1955.

33. COM, "Minutes of Second Meeting," 1.

34. Mazuzan, "Up, Up, and Away," 1153.

35. Walter Roberts to H. K. Stephenson, 4 September 1956, fol. NSF 1957, box W. O. Roberts 1940–1957 n-o, WOR/CU.

36. England, A Patron for Pure Science, 280.

37. Lomask, A Minor Miracle. Lomask devotes nine pages to the NSF's role in the creation of the National Radio Astronomy Observatory alone (139–148), but only one page on the National Center for Atmospheric Research (147) and a mere two pages on weather modification (137–138).

38. The terms weather modification and weather control are most often interchangeable. They are defined simply as intentional, human-induced changes to weather (on the short term) or climate (on the longer term).

39. Doel, "Earth Science and Geophysics," 412. For an account of some of these early efforts in the Air Force on weather modification, see Charles C. Bates and John C. Fuller, America's Weather Warriors, 1814–1985 (College Station: Texas A&M Press, 1986), 141–143.

40. Krick, trained at the California Institute of Technology, gained some fame as one of the forecasters who contributed to the accurate and important WWII D-day forecast. See Bates and Fuller, America's Weather Warriors, 94.

41. James Rodgers Fleming, "The Pathological History of Weather and Climate Modification: Three Cycles of Promise and Hype," Historical Studies in the Physical and Biological Sciences 37, no. 1 (2006): 10. A discussion of these approaches lies beyond the scope of this work, but see Fleming's article for details.

42. Chief of Bureau (Reichelderfer) to Harry Wexler, 20 February 1952, fol. General Correspondence, box 6, Harry Wexler Papers, Library of Congress, Washington, D.C. (hereafter HW/LOC).

43. Mazuzan, "Up, Up, and Away," 11–12.

44. See COM, "Informal Research Planning Conference, January 31, 1958," n.d., minutes, fol. 1, collection 8623, UCAR/NCAR Archives. Many noted in subsequent years that "NAIR" was rain spelled backwards.

45. Robert C. Toth, "Expert Doubts Reds Lead in Weather Data," n.d., orig. from Boston Daily Globe, fol. 7, J. Namias Press Clippings, 1957–1960, box 79, MC20, Namias Papers, Scripps Institution of Oceanography, La Jolla, CA.

46. McDougall, The Heavens and the Earth, 400.

47. Robert S. Divine, The Sputnik Challenge: Eisenhower's Response to the Soviet Satellite (New York: Oxford University Press, 1993), xviii.

48. COM, "Informal Research Planning Conference." For a detailed early history of AUI, see Needell, "Nuclear Reactors," 93–120. Note that HAO's incorporation (with Harvard and CU) actually preceded the creation of AUI by a few months in 1946.

49. Thomas Malone, interview by Joseph Bassi, Hartford, CT, October 29, 2007.

50. Mazuzan, "Up, Up, and Away," 1156.

51. See Appel, *Shaping Biology*, for a discussion of the various issues the NSF dealt with in formulating policies toward the funding of biological research, including "turf battles" with the National Institutes of Health.

52. McCray, *Giant Telescopes*, 37–41.

53. Mazuzan, "Up, Up, and Away," 1156.

54. "Study of Means for the Augmentation of the National Effort on Atmospheric Research," March 1958, proposal to the National Science Foundation from the Massachusetts Institute of Technology, fol. 3, collection 86231, UCAR/NCAR.

55. Waterman had served as ONR chief scientist and helped to create the broadminded policies that enabled ONR to fund many and diverse basic science research activities in the postwar United States, including HAO's solar work. For a discussion of ONR's early policies and history, and Waterman's role in formulating them, see Sapolsky, *Science in the Navy*.

56 "Conversation with Mr. Rothschild, Undersecretary of Commerce," 24 June 1957, diary note, Folder ATW Diary Notes 1957, Office of the Director Subject files, box 3, National Science Foundation, National Archives and Records Administration, College Park, MD (hereafter NSF/NARA).

57. The White House, "President Signs Act Directing National Science Foundation to Provide Research Program for Weather Modification," 12 July 1958, press release, fol. FCST-ICAS Sept 58-1960, box 45; Dr. Waterman's Subject Files, 1960–61, NSF Office of the Director General Records 1949–1963, Record Group 307, NSF/NARA.

58. David van Keuran, "Building a New Foundation for the Ocean Sciences: The National Science Foundation and Oceanography, 1951–1965," *Earth Sciences History*, 19, no. 1 (2000): 107, doi:10.17704/eshi.19.1.c531h01m58j324q6.

59. The White House, "President Signs Act."

60. Dixie Lee Ray, "A Case Study of the University Corporation for Atmospheric Research and the National Center for Atmospheric Research," 4 January 1963, special report, fol. General Correspondence, NSF, 1963, box 29, ATW/LOC.

61. Ray, "A Case Study."

62. COM, "Informal Research Planning Conference," 5.

63. Columbus O'D. Iselin to John Sievers, 13 March 1958, fol. 3, collection 8623, UCAR/NCAR. Iselin's problems with NAIR did not preclude WHOI scientists interested in the idea from participating. For example, WHOI's William S. von Arx did serve on the staff that created the Blue Book and advanced the NIAR concept.

64. "Resolution Adopted by the Governors' Conference, Fiftieth Annual Meeting, X. Meteorological Research," 21 May 1958, fol. COM 1958, COM.

65. "Preliminary Plans for a National Institute for Atmospheric Research: Second Progress Report of the University Committee for Atmospheric Research," February 1958, UCAR/NCAR (hereafter Blue Book).

66. Mazuzan, "Up, Up, and Away," 1157.

67. Blue Book, vii–viii.

68. Blue Book, 19.

69. Blue Book, 27.

70. Malone, interview by Bassi.

71. Blue Book, 22.

72. "Proposal Submitted to National Science Foundation for National Institute For Atmospheric Research," 17 March 1959, press release, Folder: NCAR 1961–62, box 65, NSF/NARA.

73. "Resolution on Atmospheric Sciences Approved by the National Science Board at Its Sixtieth Meeting," May 1959, fol. FCST-ICAS Sept 58–Dec 60, box 45, NSF/NARA.

74. Mazuzan, "Up, Up, and Away," 1158.

75. Roberts, interview by DeVorkin, 136.

76. Appel, *Shaping Biology*, 263. Appel gives a complete account of the NSF's prominent role in shaping U.S. biology in the 1940s until the mid-1970s.

77. McCray, *Giant Telescopes*, 39. See also England, *A Patron for Pure Science*, chapter 14, for a discussion of NSF's initial ventures into the world of U.S. "Big Science" activities, including such major astronomical facilities as the Kitt Peak National Observatory and National Radio Astronomical Observatory. The NSF at this time also started the ill-fated Project Mohole, an ambitious attempt to drill down to the earth's mantle.

78. England, *A Patron for Pure Science*, 147. In this, it is important to note that Waterman never wished to usurp the roles of other agencies in funding science. Rather, he simply wanted to elevate the NSF's role in shaping U.S. science. The new NSF did have the statutory authority to evaluate and correlate the scientific projects of other government agencies. Waterman, however, quickly decided these functions were beyond the reach of the NSF in practice, and he therefore decided not undertake them. As noted physicist John A. Wheeler cynically commented, the NSF did not have the political clout "to live a safe life" in the nation's capital. Wang, *In Sputnik's Shadow*, 60.

79. Congress soon removed these funds, but reinstated them after testimony by Henry Houghton on June 5, 1959. Hallgren, *The University Corporation*, 31.

80. Paul Klopsteg to Alan Waterman, 27 May 1959, fol. NCAR 1961–62, box 65, NSF/NARA. Certainly Roberts, in his mid-40s at this time, started to see full funding from governmental sources as a viable option for U.S. science, drifting significantly from his previous emphasis on the importance of considerable amounts of private funding for the health of the scientific enterprise.

81. Mazuzan, "Up, Up, and Away," 1159.

82. Minutes of Board of Trustees, UCAR, 21 October 1959, UCAR/NCAR.

83. Mazuzan, "Up, Up, and Away," n. 620.

84. Paul Klopsteg to Alan Waterman, 27 May 1959.

85. Roberts, interview by DeVorkin, 114.

86. Mazuzan, "Up, Up, and Away," 1160,

87. Hallgren, *The University Corporation*, 29.

88. Walter Roberts to Henry Houghton, 28 September 1959, Houghton Records, UCAR/NCAR.

89. Hallgren, *The University Corporation*, 34. Berkner assumed this new position in part because he had been a major advocate for satellite meteorology. He was also at this time the first chair of the newly created NAS Space Science Board.

90. "Record of Actions, Advisory Panel on Atmospheric Sciences, First Meeting," 23 September 1959, fol. NSF-APAS, box 29, AHC/LOC.

91. Mazuzan, "Up, Up, and Away," 1159.

92. Hallgren, *The University Corporation*, 31.

93. Robert G. Fleagle, *Eyewitness: Evolution of the Atmospheric Sciences* (Boston: American Meteorological Society, 2001), 86.

94. For a rather complete discussion of this aspect of meteorology in this period and how it fit into broader disciplinary issues, see Harper, "Boundaries of Research."

95. Hallgren, *The University Corporation*, 35.

96. Kassander was a meteorologist from the University of Arizona, a UCAR member.

97. For these differing accounts on Roberts's recruitment to NCAR, see Crowe et al., *Remembering Walter Roberts*, 86, and Roberts, interview by DeVorkin, 138.

98. Crowe et al., *Remembering Walter Roberts*, 86.

99. Roberts, interview by DeVorkin, 139.

100. "Quigg Newton—Memory of Walter Orr Roberts," n.d., fol. 24 "NCAR Background," box 5, QN/DPL, 2.

101. Quigg Newton to Thomas Malone, 29 December 1958, fol. 9, Admin: Memoranda-HAO, 1953–1960, John L. J. Hart Papers, UCAR/NCAR.

102. Crowe et al., *Remembering Walter Roberts*, 86–87.

103. Fleagle comments in his memoir on how the architecture of the NCAR building reflects Roberts's view on the way science should be done, and how his ideas contradicted the desires of many meteorologists for NCAR. See Fleagle, *Eyewitness*, 86.

104. Ray, "A Case Study," 55.

105. Blue Book, 24.

106. UCAR Executive Committee Meeting, 17 March 1960, minutes, UCAR/NCAR.

107. Roberts, interview by DeVorkin, 139.

108. Fleagle, *Eyewitness*, 86.

109. Malone, interview by Bassi.

Notes to Chapter 8

1. O'Mara, *Cities of Knowledge*, 5.

2. Doel, "Earth Science and Geophysics," 399.

3. "National Atmospheric Research Group Established with $500,000 NSF Contract; Walter Orr Roberts Named Director," 27 June 1960, press release, fol. NCAR 1961–62, box 65, Dr. Waterman's Subject Files, NSF/NARA.

4. Roberts, interview by DeVorkin, 139.

5. Hallgren, The *University Corporation*, 32.

6. "Report of the Evaluation Studies and Recommendation of a Site for the Headquarters of the National Center for Atmospheric Research," 9 November 1960, report, UCAR/NCAR.

7. Hallgren, *The University Corporation*, 33.

8. Roberts, interview by DeVorkin, 139.

9. Malone, interview by Bassi.

10. Alan Waterman, "Telephone Call from Senator B. Everett Jordan" 28 June 1960, diary note, fol. NCAR 1961–62, Dr. Waterman's Subject Files, box 65, NSF/NARA.

11. Malone, interview by Bassi.

12. M. A. Farrell to Hans Neuberger, February 1958, fol. 26, UCAR Correspondence, collection 850302, UCAR/NCAR.

13. Despite an apparently comprehensive archive of documents from this period and the site selection process in particular, this researcher did not find any such documents in the UCAR/NCAR collection.

14. "Report of the Evaluation Studies," 10.

15. Ray, "A Case Study," 68.

16. Roberts, interview by DeVorkin, 140.

17. "Report of the Evaluation Studies," 12.

18. Randall M. Robertson, 1 October 1960, memo, fol. NCAR 1961–62, box 65, Dr. Waterman's Subject Files, NSF/NARA.

19. J. E. Lupton to Earl Droessler, 2 November 1960, fol. NCAR 1961–62, box 65, Dr. Waterman's Subject Files, NSF/NARA.

20 "Note from Dr. Waterman," 1 November 1960, report, fol. NCAR 1961–62, box 65, Dr. Waterman's Subject Files, NSF/NARA. The November 9 site report represented the final effort after a number of reworkings in October and early November.

21. Henry Houghton, 21 October 1960, phone mem to Walter Roberts, fol. 7, box 1 of 2, Administration Division: Records, Site and Architect Selection, collection 8731, UCAR/NCAR (hereafter AD, UCAR/NCAR).

22. Henry Houghton to Alan Waterman, 28 October 1960, AD, UCAR/NCAR. None of these proposals appear to have survived in the documentary record.

23. Tician Papachristou to Walter Roberts, 28 October 1960, AD, UCAR/NCAR.

24. Mary L. Andrews, 26 October 1960, memo, AD, UCAR/NCAR.

25. Tician Papachristou, 24 October 1960, handwritten notes, fol. 20, box 2 of 2,

AD, UCAR/NCAR.

26. Walter Roberts, 24–25 October 1960, trip notes, fol. Permanent Site, 1960, box 2, Philip D. Thompson Papers, UCAR/NCAR (hereafter Thompson Papers).

27. Walter Roberts, 2 November 1960, phone memo, Thompson Papers.

28. Walter Roberts, 2 November 1960, phone memo from Earl Droessler, Thompson Papers.

29. Walter Roberts, 2 November 1960, phone memo from Henry Houghton, Thompson Papers.

30. Mary L. Andrews, 12 July 1960, memo, fol. 6, AD, UCAR/NCAR.

31. Steve McNichols to Alan Waterman, 7 October 1960, fol. 2, box 1 of 2, AD, UCAR/NCAR.

32. Alan Waterman to Steve McNichols, 22 October 1960, fol. ATW Notes, 1960, box 3, NSF/NARA.

33. Mary L. Andrews, 22 November 1960, memo, Thompson Papers.

34. This research did not find any evidence of political "blowback" or repercussions on the site selection issue in the UCAR archives, the NSF, NARA, UCAR/NCAR files, or Waterman's personal papers in the Library of Congress.

35. O'Mara, *Cities of Knowledge*, 7.

36. "Boulderites Love the City's Open Space Program, but It Took a Group of Concerned Citizens to Really Get It Rolling," *Flatirons Magazine*, Spring/Summer 1994, fol. OH 71, Janet Roberts, Carnegie Library, Boulder, CO, 32.

37. "Boulderites Love," 33.

38. *BDC*, July 1979.

39. "Boulderites Love," 34.

40. The group exists to this day, now called PLAN-Boulder County (PBC).

41. J. H. Rush, 11 October 1960, memo, fol. 7, box 1 of 2, AD, UCAR/NCAR.

42. J. H. Rush, 10 November 1960, memo, AD, UCAR/NCAR.

43. Walter Roberts, 5 September 1960, memo, AD, UCAR/NCAR.

44. J. H. Rush, 10 November 1960, memo, AD, UCAR/NCAR, 3.

45. Rush memo.

46. See chapter 3 for a discussion of the NBS laboratory move to Boulder.

47. There appears in the documentary no direct evidence on the possible orchestration of the campaign as they had done in the NBS episode, but as a minimum the Chamber played a prominent role in the NCAR effort.

48. Citizens' Committee for the NCAR Blue Line Amendment, n.d., flyer, fol. 4, NCAR Blue Line Amendment Campaign 1961, box 2, F. W. Reich Collection, CUA.

49. J. H. Rush, 2 December 1960, memo, AD, UCAR/NCAR.

50. "Open Forum Letter," n.d., fol. 4, NCAR Blue Line Amendment Campaign 1961, box 2, F. W. Reich Collection, CUA.

51. Citizens' Committee, flyer.

52. "New Dog Law Is Defeated; Blue Line, Park-Recreation Amendment Pass," *BDC*, February 1, 1961, fol. 3, NCAR, Carnegie Library, Boulder, CO.

53. "McNichols Admits He Fudged a Little," 10 March 1961, fol. 3: NCAR , BHS 328, Carnegie Library.

54. "Steve Okays $250,000 Bill for Space-Age Science Site," *Rocky Mountain News*, April 9, 1961, fol. 3: NCAR, BHS 328, Carnegie Library.

55. Leslie, "A Different Kind of Beauty," 173–221. Leslie gives a complete discussion of the selection of the architect and other building considerations.

56. Leslie, "A Different Kind of Beauty," 185.

57. I. M. Pei, 10 May 1967, comments, NCAR Building Dedication, folder 3: Dedication Speeches, Record NCAR Laboratory Dedication, box 8622, UCAR/NCAR.

58. Leslie, "A Different Kind of Beauty," 178.

59. Leslie, "A Different Kind of Beauty," 185.

60. Fleagle, *Eyewitness*, 86.

61. Leslie, "A Different Kind of Beauty," 189.

62. Leslie, "A Different Kind of Beauty," 195.

63. Doel, "The Earth Sciences," 414.

64. Leo Goldberg, interview by Owen Gingerich, October 10, 1983, transcript L831010, Niels Bohr Library & Archives, AIP, College Park, MD, 5.

65. Lewis Branscomb, interview by Joseph Bassi, La Jolla, CA, November 24, 2007.

66. Roy H. Garstang, interview by Joseph Bassi, Boulder, CO, June 12, 2007.

67. Lewis M. Branscomb, "Twenty-Fifth Anniversary of the Joint Institute for Laboratory Astrophysics" (speech at the University of Colorado, private collection, April 8, 1988).

68. Garstang, interview by Bassi.

69. Branscomb, speech, 2.

70. Garstang, interview by Bassi.

71. Branscomb, speech, 1–2; Branscomb, interview by Bassi.

72. Bartlett and Kraushaar, *A History*, 16.8.

73. Garstang, interview by Bassi.

74. Bartlett and Kraushaar, *A History*, 16.10.

75. Lécuyer, *Making Silicon Valley*, 5.

76. Joseph Kaplan to Sinclair Weeks, 29 May 1958, fol. IGY Quarterly Progress Reports, 1960, 2, AHC/LOC.

77. S. Fred Singer to L. H. Clark, 27 December 1963, fol. Space Weather Forecasting, 1963–1964, AHC, LOC (LC 19).

78. R. W. Knecht to Alan Shapley, 6 January 1964, fol. SEF 1964–1965, AHC/LOC (LC 19).

79. Alan Shapley to R. W. Knecht, 14 January 1964, fol. SEF 1964–1965, AHC/LOC (LC 19).

80. Gordon Little to Alan Shapley, 11 January 1964, fol. SEF 1964–1965, AHC/LOC (LC 19).

81. Robert H. White to Gordon Little, 15 January 1964, fol. SEF 1964–1965, AHC/LOC (LC 19).

82. J. Herbert Hollomon to Allen V. Astin, 17 January 1964, Folder: SEF 1964–1965, AHC/LOC (LC 19).

83. R. W. Knecht to J. Herbert Hollomon, 28 January 1964; J. Herbert Hollomon to Robert W. Knecht, 6 February 1964, fol. SEF 1964–1965, AHC/LOC (LC 19).

84. Doel, "The Earth Sciences," 399.

85. Doel, "The Earth Sciences," 410.

86. Bowler, *The Environmental Sciences*, 6.

87. U.S. Department of Commerce, "Dr. George S. Benton Named Director of Institutes for Environmental Sciences," 21 March 1966, flyer, G 66-60, fol. RP-4, Washington-Boulder Liaison, 1966 (Jan-Mar), AHC/LOC (LC 30).

88. Herbert F. York, interview by Joseph Bassi, La Jolla, CA, February 16, 2005.

89. U.S. Department of Commerce, flyer.

90. "A Message from Dr. White," *ESSA News*, October 4, 1965, fol. Memos from Alan H. Shapley, AHC/LOC (LC 40).

91. C. G. Little, "Draft report of ESSA Task Group on Research," 6 August 1965, memo to Robert M. While, fol. Memos from Alan H. Shapley, AHC/LOC (LC 40).

92. C. Gordon Little, memo to "All CRPL Employees," 8 October 1966, fol. Memos from Alan H. Shapley, AHC/LOC (LC 40).

Notes to Chapter 9

1. Kargon et al., "Far beyond Big Science," 335–336.

2. Lewis M. Branscomb, "Confessions of a Technophiliac" (40th Anniversary Symposium, High Altitude Observatory, private collection, October 8, 1980).

3. Markusen et al., *The Rise of the Gunbelt*, 175–176.

4. Lécuyer, *Making Silicon Valley*, 5.

5. Bruce Hevly and John M. Findlay, eds., *The Atomic West* (Seattle: University of Washington Press, 1998), 16. See also the list of attributes Paul Ceruzzi compiled from various works in his study of technology regions, *Internet Alley* (14). The first criterion on the list includes "the presence of a nearby highly ranked university." Boulder and Tyson's Corner (Virginia), the subject of Ceruzzi's study, both lack such an institution in their formative phases, but the latter's development occurred decades later than Boulder's.

6. Livingstone, *Putting Science*, 88.

7. O'Mara, *Cities of Knowledge*, 227.

8. O'Mara, *Cities of Knowledge*, 227. Much of the argument of this work centers on the importance of the university as the primary engine that powers the city of knowledge. Clearly, this research does not adequately represent what happened in the early years of science in Boulder.

9. Other factors, including weather and geography, apply as well in the development of science centers. Yet, these factors are very sensitive to individual taste and therefore tend to cancel out in the selection of science sites. Simply put, the weather

and these other environmental factors just had to be "good enough" and not tend toward extremes.

10. O'Mara, *Cities of Knowledge*, 227.

11. A detailed discussion of the exact reasons why these and many other college presidents and academics all over the nation so eagerly seized these opportunities to alter the nature of their campuses despite any perceived or actual attendant risks incumbent with government funding are beyond the scope of this work. A primary reason may center on the desire to make U.S. colleges more European-like. See Rebecca S. Lowen, *Creating the Cold War University: The Transformation of Stanford* (Berkeley: University of California Press, 1998), for a comprehensive discussion of what some consider the deleterious effects of university dependence on government funding. Others, such as Leslie in *The Cold War*, argue a similar position. Some have commented that Leslie's well-documented study of MIT and Stanford's military–industrial ties almost seem to contradict his conclusion. See Roger L. Geiger's review of Leslie's work in "Review of *The Cold War and American Science* by Stuart W. Leslie," *Technology and Culture* 35 (1994): 629–631, for such analysis.

12. If Roberts had stayed in Boulder, but Van Allen had accepted the first NCAR directorship, NCAR's location would have remained problematic. Another counterfactual posits itself here. If Van Allen had accepted UCAR's offer, might NCAR have wound up in Ames, Iowa, at the University of Iowa, where Van Allen called home? Ames, however, did not fall within the four broad U.S. regions the UCAR site committee indicated as a desirable location for NCAR.

13. Again, I am indebted to Rosenberg and Spires, previously cited, for being the first to point out the importance of sun–earth connection research to Boulder's development.

14. Lankford, *American Astronomy*, 370.

15. Lécuyer, *Making Silicon Valley*, 5.

16. Lankford, *American Astronomy*, 206.

17. See Lankford, *American Astronomy*, chapter seven in particular. See also Kevles, *The Physicists*, especially chapter 10 for a discussion of this ideology and some of its implications.

18. See Hufbauer, *Exploring the Sun*, chapter four, for an overview of space science during WWII.

19. Malone, interview by Bassi. See also Harper, "Meteorology's Struggle," for a discussion of the role of WWII in the development of U.S. meteorology.

20. Malone, interview by Bassi.

21. Walter Roberts to Terrell C. Drinkwater, 30 November 1966, fol. Damon Room (NCAR), 1 of 2, box 23, WOR/CUA.

22. The "Walter O. Roberts Hiking Trail" behind the NCAR building remains as one of the few reminders of his presence in, and importance to, Boulder. There is also a plaque dedicated to his memory at the entrance of the Pei-designed NCAR building on the mesa.

23. Scripps became part of the University of California system in 1912.

24. Rainger, "Constructing a Landscape." Rainger gives an account of Revelle's career and his effect on the La Jolla area.

25. Rainger, "Constructing a Landscape," 347. Among other ideas explicitly tying oceanographic research to national security interest, Revelle touted the atomic bomb as a "wonderful" oceanographic probe (348). One cannot envision Roberts making a similar claim on behalf of an atomic weapon's usefulness as a scientific device.

26. Indicating Revelle's strong institutional focus, he was apparently strongly disappointed when not selected as the first chancellor of the University of California's La Jolla campus (York, interview by Bassi). University of California trustees selected Herbert F. York, one of Eisenhower's science advisors and administrators, as the first chancellor of the campus—named the University of California, San Diego—in 1961.

27. See Rainger, "Constructing a Landscape" for a discussion of how Revelle's advocacy of the creation of a University of California branch campus in La Jolla arose from his work at SIO and his views on how best to structure modern research activity. General Atomic, a spin-off of San Diego–based General Dynamics, moved to La Jolla in 1956. Although a major addition to the region's technological–industrial complex, the company was not connected in any way to Scripps, Revelle, or oceanography. See George Dyson, *Project Orion: The True Story of the Atomic Spaceship* (New York: Henry Holt, 2002), 32–34.

28. O'Mara, *Cities of Knowledge*, 103.

29. Peter Galison, Bruce Hevly, and Rebecca Lowen, "Controlling the Monster: Stanford and the Growth of Physics Research, 1935–1962," in *Big Science: The Growth of Large-Scale Research*, ed. Peter Galison and Bruce Hevly (Stanford: Stanford University Press, 1992), 47–55. A rhumbatron produces stationary electric fields that can accelerate electrons, related to the klystron, a major component of the modern microwave oven.

30. See Galison et al., "Controlling the Monster," and Leslie, *The Cold War*, for discussions on the development of physics and technological research at Stanford.

31. This is not to say that Boulder was hostile to industry or practical applications of knowledge. Certainly the city in the 1960s and later became a home for "light industry" and technology, such as the Ball Brothers aerospace company and a major IBM plant. In this era of Boulder's development, the city came to look more like Silicon Valley, but this was built on the foundation of the sun–earth science of the 1940s and 1950s.

32. Paul Ceruzzi compiled a list of such criteria derived from a consensus of studies on technology regions (*Internet Alley*, 14). The ubiquitous "highly ranked research university" is among this list. O'Mara offers four major "lessons" for creating "the next Silicon Valley," including the presence of a "powerful university" (*Cities of Knowledge*, 227).

33. Markusen et al., *The Rise of the Gunbelt*, 242.

34. Livingstone, *Putting Science*, 106.

35. Kargon et al., "Far beyond Big Science," 335.

36. Tommaso Campanella, *The City of the Sun: A Poetical Dialogue*, trans. Daniel J. Donno (Berkeley: University of California Press, 1981).

37. Campanella, *The City*, 29.

38. Campanella, *The City*, 47.

Index

Abbot, Charles Greeley, 17, 25, 80–81, 100, 218n110
Advisory Committee on Weather Control (ACWC), 133, 135–36
Air Force
  Cambridge Research Center of, 60, 99, 114, 121–22, 175
  HAO funding and, 41–42, 54, 67, 78–80, 82, 98, 121–24
  IGY and, 114–15
  pointing control system and, 60–61
  see also Sacramento Peak Solar Observatory
airglow photometers, 115
Alamogordo, NM, 49–51, 206nn70, 73. *See also* Sacramento Peak Solar Observatory
Allardice, John, 68
Aller, Lawrence, 49
Allott, Gordon, 113
American Geophysical Union (AGU), 110

American Meteorological Society, 100, 132
Andrews, Mary, 154
Arnold, Henry "Hap," 45, 88
Aspen Institute, 160, 161, 224n7
Associated Universities, Inc. (AUI), 35, 38, 40, 108, 134–35
Astin, Alan, 112, 116, 162, 165
astro-geophysics, 11, 77, 97–102, 171, 218n112
atmospheric science, 125–45
  development of, in 1950s, 130–32
  NCAR's development and, 136–40
  NCAR's first directorship, 142–45
  NCAR's location and, 147–58
  NSF funding and, 134–36
  post-*Sputnik* funding and, 125–26, 128–30
  weather modification and, 128, 132–34, 226n38
Atomic Energy Commission (AEC), 197n24, 212n76

Ball Brothers Space Technology, 4, 235n31
Barker, Joseph, 82
Bartels, Julius, 217n104
Bartlett, Albert, 155, 157
Benton, George S., 166–67
Berkner, Lloyd, 120, 123
  atmospheric science and, 130–31,
    134–35, 141, 229n89
  CRPL and, 116, 117
  IGY and, 106, 107–9, 111
  "big science," 105–6, 139, 175, 219n3,
    228n77
Bjerknes, Vilhelm, 131–32
Blue Book, of UCAR, 137–38, 141, 144,
  151
Boulder, CO
  development contrasted with other
    knowledge production sites,
    4–11, 175–77
  horizontal integration of
    development of, 9, 75, 198n33
  IGY and, 108, 109–15, 122–23
  knowledge production, in 1950s,
    102–3
  knowledge production, in 21st
    century, 1–4
  limited historiography of, 11–12
  local support for scientific projects,
    10–11, 59, 61–74, 152–58
  as Marshallian district, 9, 60–61, 67,
    74, 75, 76, 157–58, 160, 162–63,
    167, 169–72
  NCAR site selection and, 148–58
  population growth in, 44, 205n37
  science and technology growth in,
    161–63
  scientific reputation of, generally,
    121, 148
  serendipitous development of, 8,
    177–78
  space environmental research and,
    163–65

World Data Centers and, 117–21
Boulder, Colorado: Development of
  a Local Scientific Community:
  1939–1960 (Rossenberg), 8
Bowen, Ira, 86, 214n44
Boyce, Joseph C., 31
Branscomb, Lewis M., 1, 161–62, 169
Breese, James L., 84
Bretz, M., 128
Brittin, Wesley, 120
Bronk, Detlev, 130, 138–39
Brookhaven National Laboratory, 35,
  38, 134–35
Brown, Francis, 118
Brown, Frederick W., 157
Broxon, James, 92
Buck, Paul H., 91–92
Bundy, McGeorge, 94–96
Burden, William A. M., 99
Bush, Vannevar, 7, 53–54, 75
Byers, Horace, 142

Campanella, Tommaso, 2, 177–78
Campbell, E. Ray, 40
Carmichael, Leonard, 86, 100, 218n110
Carnegie Institution, 32–33, 56, 81, 107
Case, Francis, 133, 136
Central Radio Propagation Laboratory
  (CRPL), 3, 57, 74–75, 78, 79, 102,
    161, 171
  formal site selection process for,
    69–72, 171
  government and local cooperation
    to bring to Boulder, 61–74
  HAO's creation and, 41
  IGY and, 110, 112, 113–15, 119
  renamed, 166
  space environmental research and,
    164–65
  World Data Centers and, 118
  world warning system and, 116–17
Ceruzzi, Paul, 5, 233n5, 235n32

Chamber of Commerce, of Boulder, 59, 66–74, 152–53, 157
Chapman, Sydney
  astro-geophysics and, 218n112
  IGY and, 106, 108, 110, 111, 119, 121
City of the Sun, The (Campanella), 177–78
Clark, L. H., 164–65
Climax Molybdenum Company (CMC), 27, 28, 29, 30, 31, 40, 82, 173
Coleman, W. C., 84
Coleman Lamp and Stove Company, 84
Colorado, University of, 3, 6, 10–11
  atmospheric science and, 125–27
  co-development with Boulder, 6, 170–71
  founding and growth of, 43–44
  HAO and, 35–42
  HAO's separation from Harvard and, 91–97
  JILA at, 161–63
  NCAR site selection and, 149
  reputation of, 101–2
  Roberts joins, 42–43
  Upper Air Laboratory of, 60–61, 102, 112
  World Data Centers and, 118
Colorado Springs, CO, 8, 27, 37, 67, 153
"Comité Spécial de l'Année Géophysique Internationale" (CSAGI), 108, 110, 117–19, 120
Commission on the Ionosphere of the International Scientific Radio Union (URSI), 108, 111
Committee on Meteorology (COM), of NAS, 130–32, 133, 134–36, 138–39, 141, 143, 153
Committee on Scientific Operations (CSO), of HAO, 40–41
  Harvard–HAO split and, 91–96

Companion to Science in the Twentieth Century (Krige and Pestre, eds.), 12
Conant, Frederick W., 99
Conant, James B., 25, 38
Condon, Edward U., 41, 92, 170
  CRPL in Boulder and, 62–70, 72, 74, 109, 210n17
  JILA and, 163
  security issues, 85
"Constituting the Postwar Earth Sciences" (Doel), 12
Cooper, Edward, 64–65
Cooperative Institute for Research in Environmental Sciences, 3
coronagraph
  funding concerns, 25–26, 29, 78, 79, 81, 201n35
  practical uses for, 22–26, 29–34
  rebuilt in Colorado, 28–29
coronagraph flare-monitoring, 113
coronal green line, 30, 33
coronal holes, 217n104
"coronavisor," 26
Craig, Richard, 99
Culver, Robert, 43

Daily Camera, 72–73
Damon, Ralph, 174
Darley, Ward, 101, 123
  HAO and, 94–98, 105, 216n84
  World Data Centers and, 119–20
data centers. *See* World Data Centers (WDCs)
Davis, Lincoln K., 19
Dellinger, J. Howard, 30, 31, 33, 41
Dennis, Michael, 7, 12
Denver, CO, 27, 83, 156
Denver, University of, 21, 37
Denver Post, 27
Department of Terrestrial Magnetism (DTM), of Carnegie Institution, 32–33, 107

DeVorkin, David, 6, 11
Doel, Ronald, 6, 11, 12
Douglas Aircraft, 99
Droessler, Earl, 123, 136, 144, 150–51,
    153–54
Dryden, Hugh, 131
dual-use projects, 7, 8, 197n26

"Earth Sciences and Geophysics"
    (Doel), 12
Echo Lake cosmic ray conference, 65,
    68
Eisenhower, Dwight D., 74–75, 88, 122,
    126, 136, 209n13, 212n76
Environmental Science Services
    Administration (ESSA), 3,
    166–67, 196n9
Epstein, Isadore, 92–93
Evans, John Jack, 29, 50–52, 60, 94,
    122–23, 206n83
Exploring the Sun (Hufbauer), 11
"extraterrestrial weather" (ETW),
    164–65

Ference, Michael Jr., 142, 143
Fleagle, Robert, 142, 229n103
Fleming, John A., 32, 107
Forman, Paul, 47
Fremont Pass, CO, 27–29, 36
Friedman, Herbert, 142
funding concerns, generally
    military-industrial-academic
        complex and, 6–8, 171, 177, 234n11
    post-*Sputnik* changes, 8, 10–12, 14,
        103, 113, 115, 117, 124–26, 128–30
    private versus military sources, 6–10
    see also specific individuals,
        organizations, and projects
Furry, Wendell, 20

Geiger, Roger, 10
Germany, 30–31

Goldberg, Leo, 10, 47, 52, 56, 92–93
Great Northern Paper Company, 82, 83
Green, Warren Kimball, 20
Greenstein, Jesse, 10
Groves, Leslie R., 205n40
Gustavson, Reuben G., 38, 39

Hagen, John P., 123
Hale, George Ellery, 161
Hallgren, Elizabeth, 12
Hamblin, Jacob Darwin, 220n23
Hanson, William, 176–77
Harper, Kristine C., 130
Harvard College Observatory (HCO)
    at Climax station, 28–34, 47–49
    Colorado locations considered for,
        26–28
    funding concerns, 23–26, 52, 57, 78
    Menzel and, 21, 90, 93, 97
    at Oak Ridge, 26, 57
    Shapley and, 21–22, 37, 50, 54–55, 78,
        87, 97, 100–101, 203n9
Harvard University
    HAO and, 35–42
    HAO separated from, 91–97
    Menzel and "Astronomical
        Geophysics" at, 101
    Roberts at, 20
Harvard–Colorado solar observatory.
    See High Altitude Observatory
    (HAO)
Haurwitz, Bernhard, 99, 140
Havice, Frank, 155, 157
Hayworth, Leland, 134
Herschel, John, 25
Herschel, William, 17, 80, 100
High Altitude Observatory (HAO), 3,
    82, 84
    Committee on Scientific Operations
        of, 40–41, 91–96
    creation of, 35–40
    fortieth anniversary, 169

funding concerns and, 41–42, 54, 67, 78–80, 82, 98, 121–24
IGY and, 112–15, 119
increase in staff size, 78
meteorological component of, 140
original board members, 40
Sac Peak and, 49–52
separation from Harvard University, 91–97
sun–weather connection and, 97–102, 217nn104,106
World Data Centers and, 118–19
see also Institute for Solar-Terrestrial Relations (ISTR)
Higman, Howard, 93, 95, 216n81
Holloman, Herbert, 165
Holly, Charles F., 43
Holme, Peter H., 40
Horner, Richard, 124
Houghton, Henry, 135, 138, 140, 148, 150–52, 228n79
House Un-American Activities Committee (HUAC). See security clearance issues
Hufbauer, Karl, 11
Hulburt, E. O., 119
Humboldt, Alexander von, 25, 80, 106–7, 220n8
Humboldtian science, 161, 166, 220n8

IGY World Warning Agency (IGYWWA), 117
Initiatory University Group, 39
Institute for Solar-Terrestrial Relations (ISTR), 97–101, 102, 119, 132, 140–41
Institute for Telecommunication Sciences and Aeronomy, 166
Institutes for Environmental Research (IES), 166–67
International Business Machines (IBM), 4, 235n31

International Geophysical Year (IGY) (1957–58), 103, 105–21, 147, 164
Boulder and, 108, 109–15, 122–23
funding concerns, 113, 221n41
origins and organization of, 106–9, 220n8
World Data Centers and, 113–14, 117–21, 222n61, 223n75
world warning system and, 115–17, 222n54
International Polar Year (IPY), 107, 109
International Union of Geodesy and Geophysics (IUGG), 108
Interservice Radio Propagation Laboratory (IRPL), 32–33, 34, 41, 57
ionosonde, 114–15, 222n46
Iselin, Columbus O., 137, 227n63

Jackson, William S., 40, 91
Jeffries, John, 122
Johnson, Edwin C. "Big Ed," 63–66, 67, 69, 70, 71, 72–74, 210nn22,23
Johnson, Lyndon, 166
Joint Institute for Laboratory Astrophysics (JILA), 4, 161–63, 169
Jordan, B. Everett, 149

Kaplan, Joseph, 109, 110–11, 119, 164
Kassander, A. Richard, 143, 144
Kemp, Frank A., 83–84
Kiepenheuer, Karl-Otto, 31
Kilgore, Harley, 53, 55, 61–62, 78
Klopsteg, Paul, 139, 140
Knecht, R. W., 164, 165
Knous, John, 65
Knowles, Thomas, 95, 96–97
Kodak, Roberts works for, 19–20
Krick, Irving, 133, 226n40

La Jolla, CA, 175–76, 235n27

Laboratory for Atmospheric and Space
    Physics (LASP), of U.S., 4
Langley, Samuel P., 17, 80
Langmuir, Irving, 133
Lécuyer, Christophe, 60, 75, 127, 198n32.
    See also Marshallian district,
    Boulder as
Leslie, Stuart, 5, 6, 159
Lippman, R. W., 84
Little, Gordon, 165
Lockyer, Norman, 17, 80, 100
London, Julius, 99, 140
Low, Robert J., 93, 95, 96, 123
Lowen, Rebecca, 5
Luton, J. E., 150–51
Lyot, Bernard F., 22–23, 26, 29

Malkus, Joanne, 137
Malone, Thomas, 138, 142, 143, 145, 149,
    174
Markusen, Ann, 5
Marshall, Alfred, 60, 198n32
Marshallian district, Boulder as, 9,
    60–61, 67, 74, 75, 76, 157–58, 160,
    162–63, 167, 169–72
Martini, Wolfgang, 31
Mazuzan, George, 12
McCray, Patrick, 10
McElroy, Neil, 128
McKelvey, Robert, 155
McNichols, Steve, 152–53, 159
McNish, Alvin G., 53, 66, 69
Menzel, Donald Howard
    astro-geophysics and, 11
    Colorado locations scouted by,
        26–28
    coronal research and, 23–26
    CRPL brought to Boulder and,
        59–60, 62–63, 65, 71, 72, 75–76
    education of, 21, 22
    funding concerns and, 35–36, 52–54,
        56–57, 77–79, 173

growing rift with Roberts, 89–91,
    216n66
HAO and, 36–37, 40–42
HAO's separation from Harvard
    and, 92–97
Roberts's education and, 17, 20
Roberts's move to Colorado and,
    28, 29
Roberts's security clearance and, 87
Sac Peak and, 45–52
security clearance issue, 86, 89–90,
    215n65
sun–earth connection and, 81
sun–weather connection and, 100,
    218n109
World War II and, 32–33
Merton, Robert, 7, 177
meteorology. *See* atmospheric science
Meyer, Erskine R., 40
Millikin, Eugene, 55
Mosley, Earl L., 40

National Academy of Sciences (NAS),
    24, 108, 174
    atmospheric science and, 126–27,
        134
    Committee on Meteorology of,
        130–32, 133, 134–36, 138–39, 141,
        143, 153
National Aeronautics and Space
    Administration (NASA), 8
National Bureau of Standards (NBS),
    3–4, 24–25, 41, 59, 120
    coronas and, 30, 31
    cryogenics lab and, 197n24, 210n22,
        212n76
    HAO and, 54, 59
    IRPL and, 32–33
    JILA and, 161
    radio station WWV, 222n54
    see also Central Radio Propagation
        Laboratory (CRPL)

National Center for Atmospheric Research (NCAR), 3, 12, 136–40
building design and, 158–59, 172, 174, 178, 234n22
first director of, 142–145
site selection for, 147–54, 160, 234n12
site selection, water and "blue line" issues, 154–58
National Geophysical Data Center, 3
National Institute of Atmospheric Research (NIAR), 134, 135, 136–37, 138, 140
National Institute of Standards and Technology (NIST), 4
National Oceanic and Atmospheric Administration (NOAA), 3
National Research Council (NRC), 108
National Science Board (NSB), 137–40, 151–54
National Science Foundation (NSF), 8, 38, 83
  atmospheric science and, 132–39, 228n78
  IGY and, 108, 113, 221n41
  literature related to, 207n90
  NCAR funding and, 134–36
  NCAR site selection and, 148, 150–51, 157
  post-Sputnik funding and, 126, 129
  proposed, 53–55, 78
  sun–weather connection and, 100–101
Needell, Allan, 106, 111
Newton, Quigg, 10–11
  atmospheric science and, 125–26
  CRPL site and, 71
  HAO and, 97, 123–24
  JILA and, 162–63
  NCAR and, 143–44, 149
Norlin, George, 44
North Atlantic Radio Warning Service, 115–17, 222n54

North Pacific Radio Warning Service, 116
Norton, K. A., 66, 69

oceanography, 110, 175–76, 220n23
O'Day, Marcus, 45–46, 60, 78
Odishaw, Hugh, 64, 96, 109, 111, 120, 123–24
Office of Naval Research (ONR), 39, 67, 79, 82, 83, 87, 98, 114, 121, 174, 175
Oldenberg, Otto, 20
Olmstead, Frederick Law Jr., 44
O'Mara, Margaret, 5, 147, 154, 235n32
Oppenheimer, J. Robert, 85
Orville, Howard T., 133, 135–36
"Orville report," 132, 133, 134, 136, 138
Owen, J. Churchill, 38, 39

Paddock, A. A., 72
Palo Alto, CA, 176–77
Palomar Observatory, 23–24
Papachristou, Tician, 151
Pecker, Jean Claude, 2
Pei, I. M., 159
Penn State University, 149, 152, 154
Penrose, Spencer, 27, 37, 203n12
People's League for Action Now (PLAN), 155–57
Physics Today, 77
Pic du Midi Observatory, 23, 26
Pickering, Edward, 24
Pietenpol, William B., 60
Plendl, Hans, 31
"pointing control" system, of V-2 coronagraph, 60–61
Polanyi, Michael, 198n33
polar expeditions, 107, 109
polarimeter, 23, 200n28

Quest magazine, 12

radio transmissions, coronas and, 30–31

Ray, Dixie Lee, 137
Reich, Francis W. "Franny," 63, 68, 70,
    72, 74, 153
Reichelderfer, Francis, 100
Research Corporation, 79, 82, 203n14
Revelle, Roger, 175–76, 235nn25–27
rhumbatron, 176–77, 235n29
Riehl, Herbert, 99
Roberts, David, 26, 29, 41–42, 78
Roberts, Janet, 28, 32, 42, 154–55, 156,
    157
Roberts, Stuart, 202n80
Roberts, Walter Orr, 17–18
    after leaving NCAR, 160, 161, 224n7
    atmospheric science and, 126, 131,
        140–42
    Boulder's development and
        knowledge production, 4–13,
        169–72
    classified nature of dissertation, 32,
        202n87
    coronal green line and, 30
    CRPL brought to Boulder and,
        59–70, 72–76
    funding concerns and, 35–36, 52–57,
        77–85, 121–24, 126, 128, 129,
        224n7, 228n80
    fundraising phases of, 172–75
    growing rift with Menzel, 89–91,
        216n66
    HAO and, 39, 40–42, 43
    as "Humboldtian scientist," 220n8
    IGY and, 110–11, 112–13, 115, 117,
        118–21
    on JILA, 163
    Menzel's security clearance and,
        89, 91
    move to Colorado, 26, 28–34
    move to University of Colorado,
        42–43
    NCAR building design and, 159
    NCAR directorship and, 142–45, 148

NCAR management and, 159–60
NCAR site and "blue line" issue, 156,
    157, 158
NCAR site selection and, 148–54
Revelle contrasted, 175–76
Sac Peak and, 45–55, 86–87, 206n76
security clearance issue, 85–91, 95
separation of HAO from Harvard
    and, 91–97, 216n83
on Shapley, 36, 201n33, 203n9
sun–weather connection, 80–85,
    97–102, 218n110
utility of science and, 18–19, 200n12
youth and education of, 19–20, 22–26
Roberts, Walter R., 89, 215n60
Robinson, Randall, 139, 150
Rockefeller Foundation, 29
Rocky Flats nuclear weapons plant, 63,
    197n24, 210n22
Roosevelt, Franklin, 75
Rose, Wickliffe, 6, 29, 61
Rosenberg, Robert, 8, 197n29
Rossby, Carl, 131–32
Rosseland, Sven, 31
Rothschild, Louis S., 130
Rush, Joseph, 157, 163
Ruttenberg, Stanley, 111–12

Sabine, Edward, 25
Sacramento Peak Solar Observatory,
    45–52, 60, 78, 90, 92, 94, 97, 172,
    206n83, 215n47, 217n97
    funding and, 54–55
    Roberts and, 45–55, 86–87, 206n76
Sawyer, Charles, 63–64, 66, 71, 72
Saxenian, AnnaLee, 9, 75, 127
Schott, Max, 27, 29
Schwabe, Heinrich, 80
Schwartzchild, Martin, 92
Science with a Vengeance (DeVorkin), 11
Science—The Endless Frontier (Bush), 75
scientific internationalism, 48, 206n61

Scripps Institute of Oceanography (SIO), 147, 175–76
Searles, Richard D., 133
Secchi, Angelo, 33
security clearance issues
  of Condon, 85
  of Menzel, 86, 89–90, 215n65
  of Roberts, 85–91, 95
  of Shapley, 85
Shane, Donald, 36
Shapiro, Ralph, 99
Shapley, Alan, 33, 61, 88, 170
  CRPL and, 41–42, 65–66, 68–70, 75–76
  IGY and, 109–12, 115–19, 121, 223n72
  space environmental research and, 164–65
Shapley, Harlow, 22, 34, 113, 172, 200n24
  coronal research and, 21–25, 36
  funding concerns and, 36, 52–56, 77, 78–79
  HAO and, 37–40
  Roberts on, 36, 201n33, 203n9
  Roberts-Menzel rift and, 90–91
  Roberts's education and, 20
  Roberts's security clearance and, 87–88
  Sac Peak and, 46–51
  security issues, 85
  spicules and, 33
  sun–weather connection and, 100–101
Singer, S. Fred, 164–65
"small science," 175
Smith, Newbern, 66, 69
Smithsonian Astrophysical Observatory, 81, 101–2, 147, 216n84, 218n110, 219n119
Smithsonian Institution, 80–81, 100
Social Studies of Science, 11–12
Solar Activity Alerts, 117
Solar Associates, 96

Solar System Astronomy in America (Doel), 11
"solar–terrestrial physics," 11
"solar–terrestrial relationships," 11
space environmental research, 163–65
Space Weather Prediction Center, of NOAA, 3–4
"Special World Interval" (SWI) warnings, 116–17
spicules, 33, 202n87
Spires, David, 12
Sputnik, restructuring of U.S. science policy after, 8, 10–12, 14, 103, 113, 115, 117, 124–26, 128–30
Stanford University, 176–77
Stearns, Robert L., 97, 205n40
  CRPL and, 65, 67
  funding concerns and, 55, 83–84
  HAO and, 37–39
  Menzel and, 87
  support for Roberts, 42–43, 90–91
  university growth and, 44–45, 205n39
Sterne, William C., 40
sun–earth connection, 11, 198n40
  before 1945, generally, 17–19
  Boulder as city of knowledge and, 102–3
  coronagraphs and Roberts's and Mendel's eventual move to Colorado, 21–34
  see also International Geophysical Year (IGY)
sun–weather connection, 25, 28, 77–78, 127, 132
  fundraising and, 80–85, 98–99
  see also astro-geophysics; atmospheric science

Teller, Edward, 131
Terman, Frederick, 177
Thomas, Albert, 151, 152
Thomas, Richard, 161–62

Thomson, Philip, 154
Tippo, Oswald, 125
Travels and Adventures of Serendipity, The (Merton), 177
Truman, Harry, 88
Turner, Frederick Jackson, 75
Turner, Robert, 156
Tuve, Merle, 56, 109, 222n46

University Committee on Atmospheric Research (UCAR), 135–40
    Blue Book of, 137–38, 141, 144, 151
    later University Corporation for Atmospheric Research, 12, 138
    NCAR staffing and, 141–45
Urey, Harold C., 55
U.S. Air Force Academy, 153
U.S. National Committee (USNC)
    data centers and, 118–21
    HAO funding, 123
    IGY and, 108–9, 111, 112, 114
    post-*Sputnik* funding and, 128
    world warning system and, 115
U.S. Weather Bureau (USWB), 130–32, 133, 164–65

V-2 research rocket. *See* Sacramento Peak Solar Observatory
Van Allen, James, 106, 107, 142, 234n12
Van Matter, B., 124
Vestine, E. H., 119–20
von Arx, William S., 227n63
von Kármán, Theodore, 45
von Neumann, John, 131

Walker, Eric, 149, 152, 154
Wallace, Henry, 25–26, 27, 173
Wallace, Henry A., 81
warning system. *See* world warning system

Warwick, John, 128
Waterman, Alan T.
    NCAR site selection and, 148, 149, 151–54
    at NSF, 54, 129, 135–37, 139, 141, 143, 228n78
    ONR and, 227n55
Waynick, Art, 152, 154
weather modification, 128, 132–34, 226n38
Webb, James E., 107, 220n11
Weeks, Sinclair, 164
Welch, S. W., 66, 69
Wexler, Henry, 131
Weyprecht, Karl, 107, 109
Wheeler, John A., 228n78
Whipple, Fred L., 57, 101
    atmospheric research of, 22, 36, 46, 47, 205n46
    HAO's separation from Harvard and, 92
    HCO and, 52, 54
    Roberts's security clearance and, 87–88
White, Robert, 165, 167
White, Wallace H. Jr., 64
Willett, Hurd, 99
Wilson, Charles E. "Engine Charlie," 122, 128
World Data Centers (WDCs), 113–14, 117–21, 222n61, 223n75
World War II
    coronagraph's practical uses in, 29–34
    science funding concerns and, 173–74
world warning system, 115–17, 222n54

Yeager, James, 72–74
York, Herbert F., 166, 235n26